Rebuilding for Resilience: A Barrier Island Case

Chamila Subasinghe

Rebuilding for Resilience: A Barrier Island Case

The Bolivar Peninsula, U.S.A.

Springer

Chamila Subasinghe
Curtin University
Perth, WA, Australia

ISBN 978-3-030-65531-0   ISBN 978-3-030-65532-7 (eBook)
https://doi.org/10.1007/978-3-030-65532-7

This Springer imprint is published by the registered company Springer Nature Switzerland AG
The registered company address is: Gewerbestrasse 11, 6330 Cham, Switzerland

For Seth and Sandamali, who ignited my resilience.

Contents

Nomenclature

ADPC	Asian Disaster Preparedness Center
ANOVA	Analysis of Variance
ASCE	American Society of Civil Engineers
BFE	Base Flood Elevation
CV	Control Value
FAO	Food and Agriculture Organization of the United Nations
FEMA	Federal Emergency Management Agency
ICT	Information and Communication Technologies
IHBS	Institute for Business & Home Safety
MD	Mean Depth
NFIP	National Flood Insurance Program
NGO	Non-Governmental Organizations
NHRAIC	Natural Hazards Research and Applications Information Center
RA	Relative Asymmetry
SIDS	Small Island Developing States
TD	Total Depth
UNDRO	United Nations Disaster Relief Organization
UNESCO	United Nations Educational, Scientific and Cultural Organization
USGS	US Geological Survey
VGA	Visibility Graph Analysis

List of Figures

List of Tables

Chapter 1
Introduction: Disasters in the Divided

This book discourses the impact of natural hazards on self-built or locally built settlements in areas where natural hazards are frequent events. The premise underlying the discourse is that the settlements under investigation are sustainable by necessity in order to survive extreme environmental events. What asserted here as "the divided" is unincorporated communities living in disaster-prone areas or environmentally sensitive ecological formations such as barrier islands (Cox & Perry, 2011). With a special reference to Bolivar Island, Texas in the USA part of the Gulf of Mexico, this investigation interrogates possible inbuilt resiliency qualities of self-built or locally built settlements on their physical makeup in respective settings.

First, this book seizes opportunities from long chronicled history of recurring disasters via exemplary 'lost lessons' to reveal lessons re-learned from each disaster cycle. Instead of altruistic advocacy of environmental conservation prevalent in the mainstream literature, it tests and contests intergenerational know-how' as an alternative or even sustainable model for progressive development for "the divided". Second, it confronts the 'global sense of place' under ground zero condition to decode reconfigured (distorted) syntax of post-disaster landscape in sustaining environmentally responsible geospatial formations. Hence, environmental responsibility asserts balancing geospatial lifecycle in ecologically formed disaster-absorbent areas similar to barrier islands. Third, the book highlights overexploitation of environmentally responsible areas and consequent incremental gentrification of unincorporated communities with tacit knowledge by transnational, commuting communities thriving on the imposing real-estate model of 'vacation homeownership'. Finally, the book offers a preamble for analysis of such gentrified settlements aftermath of natural disasters without ideologically affiliated in the kind of devastation they have finally become.

Chapter 2 draws generalizations from such settlements in the developing world as well as underdeveloped or marginalized regions of the developed world. Critical to the discourse is that these settlements at extreme opposite ends of the development spectrum have been subjected to the same socio-economic and techno-environmental constraints: "the divided", irrespective of their location in either developed or

C. Subasinghe, *Rebuilding for Resilience: A Barrier Island Case*,
https://doi.org/10.1007/978-3-030-65532-7_1

underdeveloped part of the world. Therefore, they may provide a suitable background for uncovering common factors directly responsible for their likely innate disaster resiliency. The overarching resiliency-building premise of this book hypothesises that disaster recovery demands a critical look at how extreme natural occurrences can be mitigated through knowledge that may only be available in disaster-prone settlements. Therefore, the analysis singles out stages of pre-and post-disaster recovery of spontaneous settlements in the Bolivar peninsula in order to identify various means of disaster alleviation possibilities at each stage of recovery. The final segment of this chapter includes a chronological account of Bolivar settlement development and urbanization as a port city, which situates Bolivar Peninsula in the immediate context of Galveston Island and the wider context of southeast Texas.

Chapter 2 concludes a note on an official approach to mitigation of disaster recovery among many stakeholder agencies; according to the evidence, such approaches predominantly revolve around scientific knowledge, typically paying more attention to technical issues rather than socio-economic issues. The underlying assumption for such beliefs is the notion that meticulously drafted planning regulations and building codes can adequately address disaster issues. It is not difficult to find many developing regions and underdeveloped regions of developed countries that have adopted such regulations and codes for the exact same reasons. Many federal and local governments tend to borrow planning and building codes from places where they already are in place rather than working with the local knowledge to come up with their own descriptors for codes. These prescription ordinances hypothesize that structurally sound construction materials and standardized building methods as the only means to battle extreme forces of nature. On the other hand, such prescriptions do not provide any insight into affordability potentials or cost-effective nature of extremely versatile local building methods of unincorporated communities, although time after time, they have shown certain conduciveness for a sustainable recovery. Many difficulties, including a lack of social and economic capacity, are evident in imposed regulations and codes in informal settlements in developing countries, as well as unincorporated settlements in developed countries. Considering that these settlements are occupied by more than half of the global population, one cannot exclude them not only from disaster mitigation and recovery scenarios but also from overall developmental goals. Also noted was, most of these regulations and codes demand higher levels of specialized expertise from professionals like architects, engineers, contractors, and building control officers that could not be afforded in developing or unincorporated areas. As confirmed by Ruskulis (2002), this often compromises safety in terms of structural soundness of the constructions against the intensifying nature of extremities of nature.

Communities need to be convinced of the positive impacts of planning regulations and building codes and of their necessity in hazard mitigation. Forced implementation of regulations and codes is a proven failure; people find creative ways to circumvent them. In order to increase community acceptance, polices should not alienate local planning, design and execution systems, but should complement them. Therefore, except for policies that are optimal in a given context, overly ambitious rules may not achieve the expected resiliency.

Accumulated risks are an innate part of the lives of disadvantaged groups; natural hazards are just one part of these risks. It is a fact that such communities must make choices between livelihood options for short-term existence and long-term resiliency against recurrent disasters. Also, typical livelihood and lifestyle processes they involved in, such as coastal fishing and recreation further increase their vulnerability. The notion that natural disasters are manifestations and crude reality of such vulnerabilities is becoming more apparent than ever before. Uncompromisingly, the role of humans in transforming vulnerabilities into mass disasters should seriously take into account any strategies that focus on reducing their impact. Evidently, a great majority of conventional approaches to disaster reduction react to symptoms rather than reverting to the underlying causes. Therefore, shifting the view toward mitigation among policymakers and executors is vital. Therefore, those who take stakes in disaster mitigation need to understand the nexus between development and disasters as well as its cyclical effect on accelerating vulnerabilities.

Past natural disasters have provided ample evidence to support the idea that financing pre-disaster mitigation is more economical as well as less stressful than relief and reconstruction; this is based on the notion that mitigation is not a choice but a necessary part of any developmental schema. Mitigation is more about wisdom of expenditure of funding rather than the monetary value of funding allocation. Top-down methods with heavy administrative structures often promote dependence instead of self-reliance and may even contribute to increased vulnerability under post-disaster stressors. As they are more case sensitive and thus informed, site-based mitigation measures often consume fewer resources compared to pre-packaged systems with generalized disaster management protocols.

Chapter 3 reviews literature related to sustainable recovery following natural disasters, with a specific focus on environmental sustainability as one of the key contributors toward resiliency. Issues central to the discourse emphasize redefining post-disaster recovery as a collective choice that communities and external agencies make under pre-disaster conditions prior to an extreme natural event. This chapter further reviews the existing knowledge base pertaining to spatial aspects of disaster-prone settlements that utilize local/traditional/indigenous knowledge or "know-how" as an integral part of achieving sustainability in pre-disaster mitigation, during-disaster management, and post-disaster reconstruction. Furthermore, the chapter delves into assessing community-based knowledge for modern wisdom to face changing modes of recurring disasters and the possibilities of introducing such wisdom to localities challenged by infrequent disasters will be investigated. Next, real-world examples examined to demonstrate how different aspects of community-based expertise or "know-how" have been developed and adapted to mitigate and withstand the aftermath of natural disasters across different demographic divisions. The final portion of the review focuses exclusively on the post-Hurricane Ike disaster evaluation and the possibility of adapting spatial analysis to reveal the built-in resiliency of Bolivar neighbourhoods against hurricanes. Because of the informal cum unincorporated nature of settlements on the Bolivar Peninsula, the cases investigated have crossed the boundaries from spontaneous to vernacular. The term vernacular has been used as an alternative to the term spontaneous in several segments of this chapter to identify methods of construction that use available local resources and traditions to address local needs and circumstances. This is on the contested understanding that vernacular architecture tends to evolve over time to reflect the environmental, cultural, and historical context in which it exists. However, it has often been dismissed as crude and stagnated but also has proponents who highlight its "role disregarded" in increasing vulnerabilities.

The literature reviewed in Chapter 4 supports the existence both of local knowledge for resiliency and survival strategies, particularly for self-built and locally built settlements. It further suggests that location and configuration of such settlements and their spatial make-up of individual dwelling units do constitute an ability to respond to adverse climatic conditions. Although it was not conclusive, many researchers found that the severity and type of damage to a particular structure depended on several factors, including the quality of original construction, elevation above mean sea level, and proximity to the shore. Furthermore, elevation has been depicted as the most critical determinant for surge-associated damage. Conclusions drawn from the literature review indicate that architects, engineers, and planners tend to view the vulnerabilities of people and the environment as the result of weak designs and poor constructions. Such points of view take on the technical task of fixing the offending buildings and community design. The apparent hypothesis behind this argument involves preventing future disasters by adapting expensive state-of-the-art designs, robust materials, and cutting-edge construction details. Furthermore, building physicists have improved the accuracy and precision of estimates of ground shaking for microzonation and storm surge heights to the nearest centimeter. Increasingly detailed, performance-based building codes have been produced as pioneering professions have pursued their adoption as public policy decisions in many parts of the world. Most of the efforts have focused on changing the structures while less attention has been paid to effecting structural change needed within specific social, political, cultural, and economic environments to enable local knowledge. Consequently, the groups who intended to be the beneficiaries of these advances in both technical knowledge and policies have sometimes become steadily more vulnerable due to the autocratic nature of the process. This suggests the necessity for a community-based approach to reconstruction for disaster risk reduction. Noticeably, some examples of community-based mitigation are derived as lessons; such awareness comes from learning from the past, building relations with communities, encouraging participation, involving local builders and artisans, building local capacity, documenting and sharing accumulated knowledge, and most importantly, influencing formal education.

Scaffolded on the previous chapters, Chapter 4 demonstrates a theory-testing model to extend the discourse beyond status quo. It processes ways of synthesising theories and approaches conducive to the central thesis of grounding long-term disaster resiliency through sustainability goals. This is to be understood as an empirical relationship between spatial integration and disaster resiliency as portrayed in hurricane-prone spontaneous settlements in Bolivar Peninsula, Texas in some form of tacit knowledge. The chapter explores spheres of sustainability in terms of their innate disaster survival of coastal settlements that were built by small-scale local contractors, sometimes identified as carpenters, artisans, or homebuilders. The selection of settlements with specific types of dwellings was based on the understanding that local technology and skills best describe responses to local environmental conditions, including protections against extreme natural events specific to the area. This chapter further examines the survival strategies of coastal settlements for surge or flood surge exposure in regard to their permeability of residential typologies and spontaneity of settlement organization. This is based on the premise that speedy recovery during the post-disaster phase is based upon many locally manageable factors including readily available local expertise, infrastructure, and local materials that are amply available at an affordable cost for both owners and donors. The methodology employed includes syntactical methods to decode spatial patterns connected with surge exposure. One such pattern, the visibility or axial maps, demonstrates integration, permeability, and connectivity among settlements. Second, based on these spatial maps, the characteristics of the surviving residential built fabric reveal performance of architectural and structural make-up on a permeability rubric. The key query finalized to lead the discourse has been handled within the broader context of different causes and types of hurricane damage and thus, detailed in the final portion of the chapter.

A hybrid of field query tactics certainly can unveil hidden constructs of the socio-cultural production of place as well as space discoursed in Chapter 3. Some of these may not be completely latent, but most are interpreted out of context including their operational meaning in disaster discourse. This is an inherent condition of research, recovery, and mitigation of disaster and mass emergencies, perhaps, due to their frequency, magnitude and humanitarian cause. The chapter seizes space accumulation of progressive settlements as well as the architectural makeup of building elements from the smallest building blocks of such settlements: detached family dwelling. The goal was to broaden the knowledge base of disaster resiliency of a particular intact spatial unit from planning of settlements to designing of individual dwellings. While syntactical analysis furnishes an integrated map of spatial connectivity among open space networks of the settlements, the permeability rubric reveals the structural patterns of elevation among residential units. However, it was revealed that analysis of elevation affecting permeability should not be investigated in settlements that show no or insignificant correlation between settlement connectivity and surge damage. In essence, object-specific studies should only be carried out in settlements where spatial connectivity informs aspects of surge permeability in regard to ground elevation of individual properties. Therefore, the need arises out of necessity in assigning credibility to elevation characteristics on surge permeability that emphasized in the disaster literature for such surge-resilient houses. It may further validate the investigation by finding the degree to which those characteristics networked with the grid of open spaces are accountable for controlling permeability.

Although this discourse is a spatial study that narrowly focuses on the syntactical analysis of the Bolivar open space network, the analytical tool Space Syntax only furnishes connectivity measures for axial segments in the Bolivar neighborhoods. However, syntactical analyses employed in Chapter 4 best represent settlement formations as a foundation for near accurate representation of connectivity within the sampled settlements. The results and discussion accompany an introduction to the nature of the damage following Hurricane Ike, along with a detailed description of the cohort of settlements and their axial dynamics. The physical attributes of the Bolivar settlements are presented from the point of view of axial integration and the nature of the damage. An inevitable comparison on nature of surge damage is also extended to Bolivar Peninsula's neighbor, Galveston Island. The visibility maps further demonstrate different levels of physical connectivity among linear developments of the gridiron patterned arteries and associated open space networks in different parts of the Peninsula. Jointly, the twofold data analysis employs a study of hurricane damage of the studied settlements at the neighborhood scale and of individual dwelling units at the axial line scale. The use of syntactical measures to interpret spatial transactions between physical connectivity and various levels of hurricane damage forms the major portion of the chapter. This is followed by an establishment of the statistical method and tools and the theoretical underpinnings of both an Analysis of Variance and a Logistic Regression to unveil the possible relationships hypothesized in the key query. The final segment of the results and discussions analyzes damage to individual properties based on their connectivity and the damage index. Residential properties along the most connected open space network in one of the highly connected Bolivar neighborhoods at Port Bolivar have been investigated using a logistical regression on data obtained through a permeability rubric. Logistic Regression is the statistical tool used to interpret significance portrayed in the rubric, based on binary data (*no damage* or *destroyed*) and on the categorical nature of the permeability characteristics. The conclusionary remarks of the chapter reestablish why these twofold statistical analyses are critical to build a bridge between progressive developmental patterns of spontaneous settlements and surge vulnerability.

Chapter 5 draws on magnitudinal and historical trajectories of Hurricane Ike as it is being considered one of the most devastating hurricanes to have made landfall on the Gulf of Mexico, particularly Bolivar Peninsula. The catastrophe offered a tabula rasa for hurricanes vs. settlements studies. First, it easily calibrated the magnitude of the destruction in the Bolivar Peninsula into a big data set due to its clearly defined geospatial boundaries. Second, it brought to light the latent spatial underpinnings of transactions between natural events and built fabric at both the settlement and dwelling levels. Building on this, the chapter reports on the threefold investigation criteria: (a) simple fluid dynamics that authorizes behavior of surge flow from high to low-pressure zones (b) depth, connectivity, and the layout pattern of open spaces that alter surge flow, and (c) built mass, ground elevation, and structural integrity of the properties that manipulate surge flow. In the process, this inquiry extended its discourse to reveal a possible relationship between the physical connectivity among the settlements and surge damage to dwellings within the studied settlements. This adopted the established notion that uninterrupted escape routes and options connecting various parts of settlements within (local streets) and outside (global streets) as one of the contributors dealing with surge permeability. Furthermore, for each dwelling studied, the connectivity was averaged based on the number of global and local streets that directly networked to form a permeability measure. The mean connectivity was assessed against a five-fold damage scale: none, minor, moderate, major, and destroyed. Statistical inference was employed to reveal if one given variable had a significant effect on a second variable across any of the groups under study; this was employed to find whether the hurricane damage is differentially expressed in at least one of the groups analyzed. The evidence supported the theory that the structural elevation of houses above Base Flood Elevation was critical to surge permeability. Hence, via structured observations, in-depth building biopsies examined elevation characteristics of survived houses under five planning, architectural and structural typologies: ground orientation, ground elevation, ground enclosure, external bracings, and external attachments. The analysis of a rubric formed from these group-characteristics revealed that typologies or combinations of built-characteristics best represent the architectural make-up of the surviving structures.

Because the analysis only considered unobstructed routes of surge into account, its reliability depends on the accurate representation of the access grid of settlements being analyzed. Property damage levels are categorical and damage assessment must be consistent through pre-identified characteristics assigned to each damage level. As the survival data was binary (survived and destroyed), the estimation of survival probabilities of houses was also based on precise estimation of each damage level. Although the dependence on the five groups of structural elevation characteristics was high, they were distilled from existing research evidence. After an indefinite time, the rigor of disaster data decreased and access to data increased. This was due to homeowners' immediate initiatives to make their damaged homes habitable, which critically affected the rigor of the field data.

A nexus between connectivity of open space network and hurricane damage from surge does exist; flood escape routes do get affected by built masses and specific spatial and built characteristics of settlements and houses, which are their building blocks. Typical gridiron patterned road arteries established by early Spanish settlers still plays a significant role in the intercostal transportation system and in the local and global movement systems of the peninsula. While *no damage* and *destroyed* damage levels directly deal with settlement syntactic analysis, the *minor*, *moderate,* and *major* damage levels suffer from small sample sizes. The only significant groups in the permeability rubric, ground elevation and enclosure, also exhibit a similar statistical relation to survival binaries. The high standard error for one of the ground enclosure characteristics (exposed plinth), makes statistical significance for its combinations with ground enclosure inconclusive. However, given the high number of variations between the results used to calculate the mean value for one of the samples (due to the small sample size for ground enclosure two), the mutual impact between a group and group characteristics have not been discarded.

Chapter 6 presents a meticulously unbundled set of deployable insights with specific merits to policy and practice of built environment at the grassroots level. What central to the conclusions presented in this chapter on the spatial-aspatial dialectic among different abodes in spontaneous settlements is the dynamic community response to extreme natural events. Therefore, an objective evaluation of such responses could be conducive to the processes of grounding complex concepts characterised in abstract yet experiential idioms of spontaneous settlements as well as their responses to adverse environmental conditions. Thus, the degree of precision does not depend solely on the processes and procedures, but also on the basic assumptions by which such processes and procedures are assembled. Therefore, these conclusions should not be looked at as an antidote for a phenomenon that is ever-changing yet offers a valuable set of diverse opportunities. The conclusions affirm that people do know how to build their abodes to compliment not only their socio-cultural environment but also to make the best out of the climatic conditions in places where they live. However, they often compromise in planning, designing, and constructing their settlements and abodes due to several economic and non-economic reasons. Economic reasons are more obvious than non-economic reasons such as statutory push to receive acceptance and adherence into mainstream planning and building framework.

The conclusions in the final chapter support generalizability over spontaneous settlements in environmentally sensitive locales, especially for coastal landforms similar to barrier islands. However, the general policy implications may have a broader application for rethinking development and revisiting sustainability for any coastal development associated with spontaneous settlements. Moreover, the utility implications of general conclusions distilled architectural make-up of surviving Bolivar houses into a cohort of both physical and functional characteristics. However, these characteristics may not be built typologies but representative of residential models of survived settlements. Due to the quantitative nature of the syntactical method and the statistical analysis used in the analysis, it would be easy to use large samples to provide clearer statistical significance in a follow-up study. Therefore, the investigation emphasized the measures of reliability over validity.

The grand summary of conclusions focus on three areas of spatial intelligence: first, on the need for developing a system to evaluate spatial attributes of rural spontaneous settlements with subtle urban characteristics. Second, the importance of assigning objective values to the socio-cultural production of space and place. Third, in achieving resiliency via sustainability, the necessity of capitalizing on local disaster coping strategies that required to establish long-term resiliency among vulnerable communities. Potentially, the outcomes of this discourse can support three-fold significance critical to sustainable disaster recovery: preserving place character, social justice among affected groups, and rapid recovery. First, sustainable rebuilding not only reestablishes pre-disaster socio-cultural and economic barriers but also highlight preservation of character of settlements that is essential to regain the cohesive community identity to face future disasters. Second, social justice perhaps is possible via levelling of prevailing social disparities by accepting and integrating resiliency characteristics that may present in unincorporated communities toward positive transformation among different social strata. Third, the rapid yet steady recovery leading to the return to normalcy may achieve long-term resiliency against recurring natural hazards. Therefore, one of the most critical factors of post-disaster recovery, funding for speedy rebuilding process, is significantly supported by manageable damage as well as easily replaceable, regionally available, and locally understood resources.

References

Cox, R. S., & Perry, K.-M.E. (2011). Like a fish out of water: Reconsidering disaster recovery and the role of place and social capital in community disaster resilience. *American Journal of Community Psychology, 48*(3–4), 395–411. https://doi.org/10.1007/s10464-011-9427-0

Ruskulis, O. (2002, June). Developing processes for improving disaster mitigation of the urban poor. *Basin News, 23,* 8–10.

Chapter 2
Global to Glocal

> There are two big forces at work, external and internal. We have very little control over external forces such as tornadoes, earthquakes, floods, disasters, illness, and pain. What really matters is the internal force. How do I respond to those disasters? Over that, I have complete control. (Buscaglia, 1983, p. 171)

The sheer magnitude of catastrophic natural events has increased the awareness of disaster response and vulnerability of settlements. A few such disasters, especially the 2004 Indian Ocean Tsunami and Hurricanes Katrina and Rita in 2005, raised the bar for post disaster management, including rebuilding. Several communities affected by such events have yet to be restored to their pre-disaster conditions, and some populations have been permanently displaced. Because of the current financial environment and the rising number of natural disasters, the funding formula for rebuilding often favors cutting costs as much as possible. On the other hand, some communities have shown a remarkable ability to recover from periodic or recurring disasters with minimum or no external help. Many people living in hazard-prone areas have demonstrated specific knowledge and skills that confirmed their ability not only to withstand extreme climatic conditions, but also to bounce back to normalcy through rebuilding. Notably, this ability is strong for indigenous communities living in spontaneous or vernacular settlements under continuous threats from natural disasters.

Communities that are highly vulnerable to natural disasters usually recover slowly, but communities that have developed less vulnerability to disasters often recover and resume pre-disaster routines relatively fast. Therefore, the key concepts and strategies of this book focus reducing both vulnerability and recovery time for higher resiliency via sustainability against natural disasters (see Fig. 2.1). It further proposes integration of local knowledge from less vulnerable settlements to mitigate, prepare, and plan for the aftermath of natural disasters with special regard to rebuilding homes. From its examination of a similar case, this analysis found that hurricane-prone spontaneous settlements on Bolivar Peninsula embody a certain compendium of planning and building practices that sustained a unique set of built-characteristics through numerous hurricanes.

C. Subasinghe, *Rebuilding for Resilience: A Barrier Island Case*,
https://doi.org/10.1007/978-3-030-65532-7_2

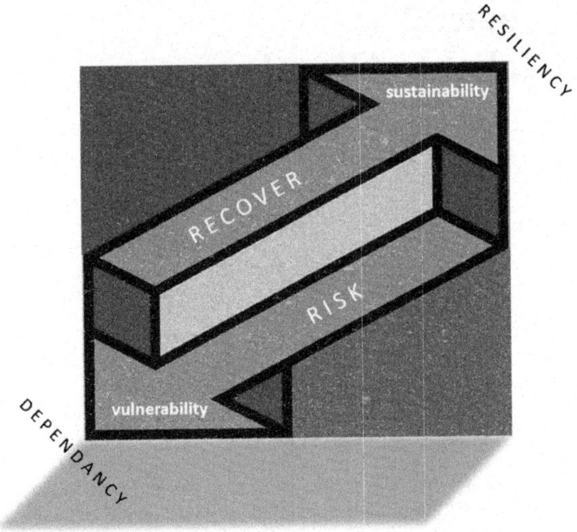

Fig. 2.1 The status quo of lessening vulnerability and speeding recovery for sustainability: resiliency

Disaster Epidemiology

Since the 1960s, the number of reported natural disasters has increased dramatically. The perceived increase in the rate of natural hazards could be a result of a combination of instant and extensive reporting by mass media, state-of-the-art disaster recording and measuring technology, and most importantly, occupation or over-occupation of geological formations meant for natural hazard absorption. Does this mean there more natural disasters are actually occurring, or simply that improved communications mean more disasters are being reported? It is highly likely that part of the increase is due to more events being reported, even relatively insignificant events. However, the number of significant natural disasters such as floods also appears to have increased. In addition, more people now live, work, and build in places considered as hazard-prone. An improved global and comprehensive tracking network, especially for earthquakes, climate change, and other factors such as topsoil erosion and deforestation (in the case of floods), has also contributed to increased reporting of natural disasters (see Fig. 2.2).

The Global Perspective

The increase in reported natural disasters is especially noticeable in coastal areas because these areas are particularly vulnerable to floods, cyclones, hurricanes, and tidal waves. These coastal belts have also experienced some of the greatest population growth since the 1960s. Many of these same coastal areas have also experienced an increase in development and construction, including some areas where development

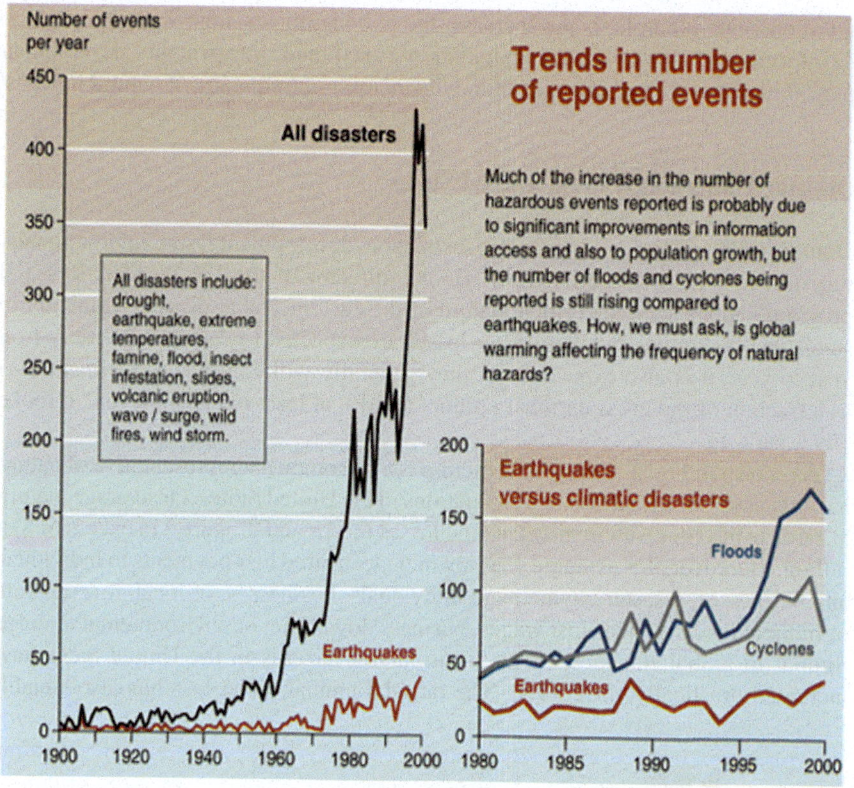

Fig. 2.2 Natural hazard occurrences have almost doubled in the last 20 years (*Source* Centre for Research on the Epidemiology of Disasters [CRED], 2009)

was previously banned because of environmental concerns. Such increases in population and development have led to situations where naturally occurring hazards are more likely to turn into natural disasters. Not only the rapid growth of world population drastically alters the number of natural hazard occurrences, but also triggers development activities in the areas where previously constructions were restricted due to their ecological vitality. In addition, introduction of extremely environmentally sensitive technologies to regions that are disaster-prone make them increasingly vulnerable to extreme natural events. Rapid concentration of economic activity in large coastal cities and industrial areas also increases the potential for catastrophes. Finally, the environmental changes caused by man have already increased the danger of catastrophic natural disasters in many locations around the world (Berz et al., 2001; Hillmann & Ziegelmayer, 2016; Zickgraf, Telenti, Sara, Ozer, & Gemenne, 2016).

The nature of any future developments in urban areas is such that often the only land available for urban growth is in disaster-prone areas such as flood plains or steep slopes with the potential for landslides. According to United Nations Environmental Program (UNEP) statistics in Fig. 2.1, this drastic increase in disasters raises vital questions. For example, is the increase due to a significant improvement in access to information? What part does population growth and infrastructure development play? Finally, is climate change behind the increasing frequency of natural hazards?

Disaster Vulnerability in the United States

Domestically, natural disasters are a sustainable development issue; internationally, they threaten global security. H.W. Hooke, director of the US Weather Research program, emphasized the critical nature of the current global disaster vulnerability status. He stated that "while disaster losses are expected to vary considerably from year to year, it is also expected that they generally will continue to increase, even as a fraction of the gross national product (GNP), at least in the short run" (Hooke, 1997, para. 1).

As shown in Fig. 2.3, the global picture is also somewhat representative of natural hazard occurrences and damage estimations in the United States. On average, natural hazards in the US result in direct annual losses (crops and property) in excess of \$7.6 billion. Moreover, this estimate does not include insured loss payments to individuals and businesses, disaster payments to individuals, businesses, or local governments, or indirect losses such as lost wages, business downtime, or environmental damage. Similar to global statistics, losses from natural hazards in the United Sates have increased drastically since 1960. The decadal annual mean loss has also steadily

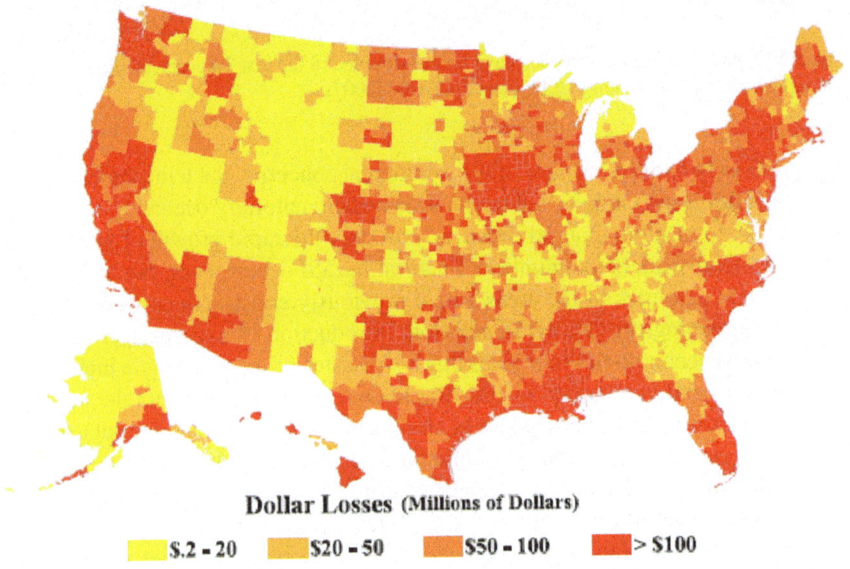

Fig. 2.3 Total property and crop losses, 1960–2003 (adjusted to 2004\$), mapped by quartiles (*Source* Spatial Hazard Events and Losses Database for the United States [SHELDUS], 2011)

increased since the 1960s, peaking at $14.4 billion in the 1990s. Even considering data for just the first four years from the present decade, it appears that the nation is on track to surpass the 1990s average annual loss (see Fig. 2.3). In addition to being the most costly decade for natural hazards in United States history, the 1990s were the most deadly, with more than 5200 fatalities attributed to natural hazard events (Cutter & Emrich, 2005).

Both governmental and nongovernmental organizations have expressed concerns over financing recovery projects during the current economic downturn. It is well known that the main benefactors of disaster recovery, including the United States, are in a major recession; therefore, it is obvious that allocations available to finance disaster recovery have become much tighter (Alexander, 2016; Baily & Elliot, 2009).

One may argue that these occasional peak periods of disaster losses are due to singular large events, such as hurricanes or earthquakes. However, such singular events alone cannot provide a holistic understanding of the drastic increase in losses due to natural hazards. The statistics shown in Fig. 2.4 provide evidence of higher dollar losses from weather-related events versus less dollar losses from other general hazards. One explanation for this difference in loss amounts is a higher frequency of weather-related disaster events. By comparison, occurrences of other geophysical events are relatively infrequent and the resulting damage losses are significantly less than weather-related losses. In certain types of weather events, losses have actually been reduced; this may due to advancements in Information and Communication Technologies (ICT) including detection, early warning systems, and mitigation such as flood- proofing, hurricane shutters, and tornado safe rooms (Cutter & Emrich, 2005; Toya & Skidmore, 2015). It is not difficult to conclude that increased disaster

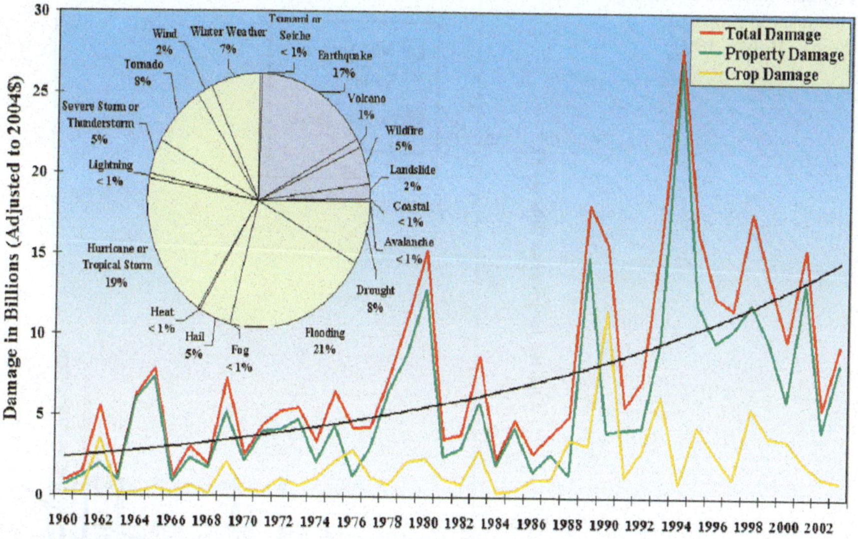

Fig. 2.4 Natural hazards losses from 1960–2003 based on the spatial hazard events and losses (*Source* SHELDUS, 2011)

losses occur because more people and property are now located in disaster-prone areas. While hazard events fluctuate exponentially, our understanding of them also improves. However, the frequency of these events remains relatively constant (Board on Natural Disasters, 1999; Van der Wink et al., 1998).

More and more areas of the United States are becoming increasingly vulnerable to disasters. In addition, people continue to live and build in areas that should be reserved as disaster absorption zones including barrier islands, marshes, low-lying and water-lodging areas, and mangroves. Furthermore, losses from uninsured property damage are on the rise because of the high cost of property insurance in these areas (Kunreuther, 2015; Kesavan & Swamminathan, 2006) (see Fig. 2.5).

The dichotomy between rebuilding after each periodic disaster and the funding agencies that are compelled to support cyclical rebuilding efforts provides an appropriate model for sustainability recovery. The World Bank, one of the major funding agencies in reconstruction, reported that two of the ten most frequent negative outcomes of World Bank disaster projects are "subsequent disaster lessened the project impact" and "research component not undertaken/implemented" (Independent Evaluation Group (IEG), n.d, p. 7) (shown in Fig. 2.6).

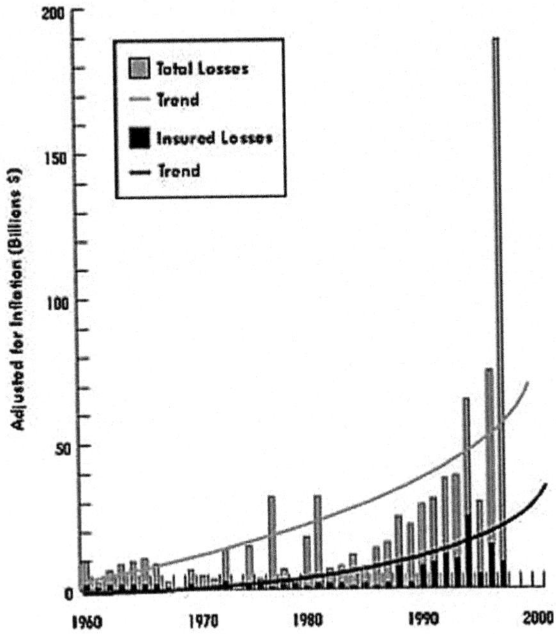

Fig. 2.5 Increase of damage to uninsured properties (*Source* Hooke, 1997)

10 Most Frequent Negative Outcomes of World Bank Disaster Projects

1. Subsequent disaster lessened the project's impact	**73**
2. Unsuccessful mitigation activities	32
3. Infrastructure reconstruction was not completed / not successful	28
4. Problems with procurement had a negative impact	27
5. Lack of maintenance lessened the project's impact	**22**
6. Shortfall in counterpart funding	22
7. Conceptual failure during design	19
8. Objectives not attained due to reallocation	18
9. Research component not undertaken / not implemented	**17**
10. Problems with distribution of resources, goods, or services	**13**

Fig. 2.6 Ten most frequent negative outcomes of World Bank disaster projects (*Source* IEG, 2005. *Note* Facts related to post disaster rebuilding highlighted)

Sustainability and Disaster Vulnerability

In order to be cost-effective, hazard reduction requires a diverse, interconnected range of actions, including: (a) comprehensive hazard identification and risk assessment; (b) wiser land use; (c) improved structural design; (d) improved building codes and practices; (e) greater public awareness, education, and training; (f) improved predictions and emergency response; and (g) more effective relief and recovery (Hooke, 1997). Even though the magnitude and frequency of disasters have increased drastically over the last twenty years, the capital expenditures for disaster reconstruction have decreased for several reasons, especially because of recessions in major donor countries (see Fig. 2.7).

Sustainable disaster recovery challenges any community, including spontaneous settlements. Apparently, recovery is more difficult for self-generated settlements such as spontaneous settlements because most of them do not fully confirm to land use patterns accepted by federal and state authorities. Spontaneous settlements, like any other human settlements are formed and shaped in the sense that purposeful changes are made to the physical environments through a series of choices among the alternatives available (Mitchell, Enemark, & van der Molen 2015; Patton, 1988). Most importantly, rebuilding of spontaneous or vernacular settlements as a part of long-term hurricane recovery is among the least explored phases of disaster. Identifying the root causes of physical damage as well as disaster resiliency characteristics is pertinent to the reduction of future recovery complications.

Accomplishing sustainability has become an essential part of any study that deals with resource management, whether concerns business, economic, tourism, cultural, or disaster recovery. Even though the notion of sustainability is multifaceted, it

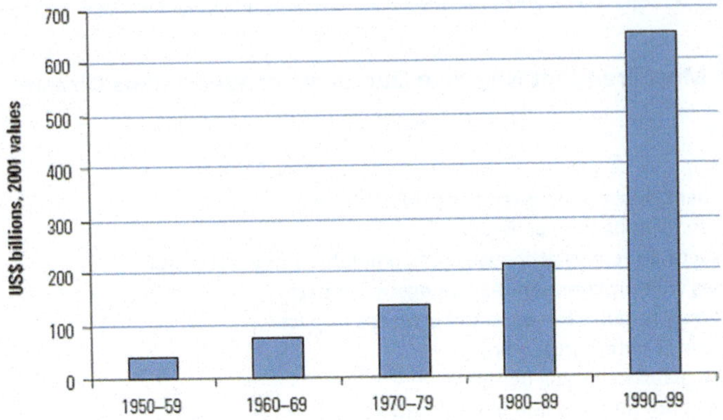

Fig. 2.7 Increase of damages from natural hazards (*Source* IEG, 2005)

fundamentally recognizes three vital notions such as environment, economy, and community. These must be considered as a whole in any reconstruction and recovery effort and none of the efforts should be compromised (Al-Nammari, 2006; Eid & El-adaway, 2016; Berke & Beatley, 1997; Daher, 2000; Giddings, 2000; Hardy & Beeton, 2001; Lefevre, 2000; Mileti, 1999; NHRAIC, 2001; Rothrock, 2000).

The 1987 Brundtland Commission report, *Our Common Future*, defines sustainable development as:

> … development that meets the needs of the present without compromising the ability of future generations to meet their own needs. It contains within it two key concepts:

- the concept of 'needs', in particular the essential needs of the world's poor, to which overriding priority should be given; and
- the idea of limitations imposed by the state of technology and social organization on the environment's ability to meet present and future needs (p. 43).

A working definition of sustainability is essential to establish a connection between disaster recovery and reconstruction. Many argue that the idea of sustainability was unfamiliar to our ancestors simply because it was simply their life style and they did not know any other way to live their lives rather than being sustainable. Therefore, they did not look for a definition of the term. Sustainability was simply the responsible use and development of materials and skills. Figure 2.8 illustrates the socio-economic counterparts of environmental sustainability.

Because it is multifaceted, sustainability calls for social, environmental, and cultural completeness. The social facet calls for social equity, justice, and equilibrium in distribution of resources, while the environmental facet calls for sound strategies that avoid depletion of environmental resources. The cultural facet includes maintaining cultural resources for future generations and reducing losses due to carelessness, damage, or disasters (Jigyasu, 2015; Giddings, Hopwood, & O'Brien, 2002). Moreover, sustainability provides a framework for understanding disaster recovery and reconstruction, as well as providing a common ground for understanding the

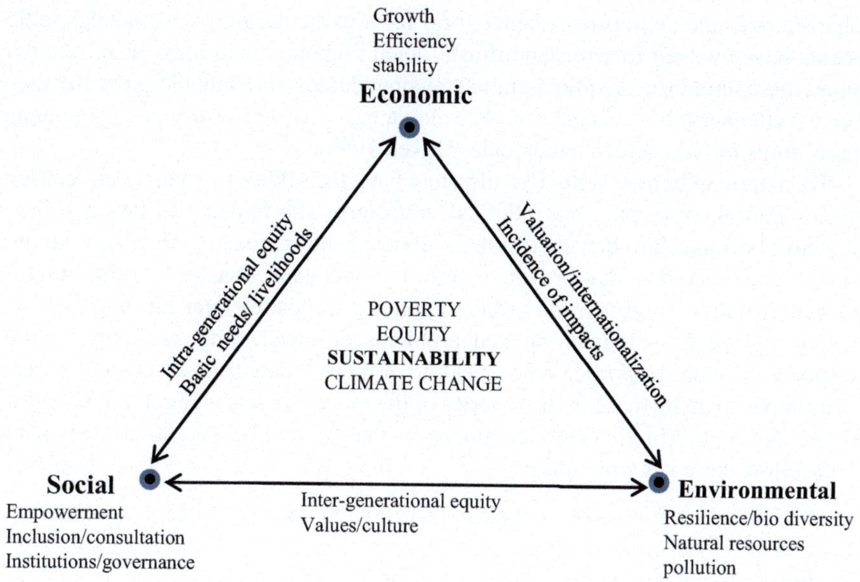

Fig. 2.8 Socio-environmentally bearable, socio-economically equitable, and environment-economically viable systems provide resilience against natural hazards over time (*Source* Munasinghe, 1994)

disaster resiliency of vernacular settlements, reconstruction after disasters, and long-term and short-term recovery.

The post-disaster period demonstrates four main operational phases of recovery. These include the emergency period (response), the restoration period (short-term recovery), the reconstruction one period (long-term recovery), and the reconstruction two period (commemorative) (Berke & Beatley, 1997; Haas, Kates, & Bowden, 1978). However, for the purpose of this research study, the reconstruction phase is referred to as the time needed to repair a dwelling or a settlement after a hurricane. Disaster recovery and rebuilding take place in a sequence of four overlapping periods: emergency, restoration, reconstruction to pre-disaster levels, and betterment reconstruction, which may entail improving infrastructure and civic spaces (Kates, Colten, Laska, & Leatherman, 2006).

Approaches to Post Disaster Sustainability

In order to comprehend the key approaches or frameworks, and the principles and indicators guiding the development of post-disaster sustainability, it is extremely critical to find where disaster resiliency sits in the domain of sustainability for spontaneous settlements. Because spontaneous or vernacular architecture by definition characterizes the appropriate shelter form for a particular place, vernacular settlements have evolved to withstand drastic local climatic conditions. In most situations, these settlements require minimal repair under locally available expertise; these repairs after periodic disasters complete maintenance as well as further strengthening procedures for vernacular settlements (Oliver, 1998).

By nature, extreme events like disasters have the ability to change and confuse socio-cultural, economic, and political indicators. The boundaries between these indictors become blurred, which allows a more holistic approach to analyzing circumstances and situations that provide insight to understand disaster housing in terms of sustainability. However, the present approach to post-disaster housing does not respond properly to the interrelated problems of speed, time, economy, and no response or under-response. Whenever one of these problems is addressed alone, it contributes to an increase in the severity of the other problems (Ehrenkrantz, 1989).

According to Mileti (1999), communities that want to become more sustainable will follow these key principles:

a. Maintain and, if possible, enhance, its residents' quality of life.
b. Enhance local economic vitality.
c. Ensure social and intergenerational equity.
d. Maintain and, if possible, enhance, environmental quality.
e. Incorporate disaster resilience and mitigation.
f. Use a consensus-building, participatory process when making decisions.

In order to survive disasters, to maintain their special character (vernacular or indigenous quality), and continue to be habitable places for their members, communities require mechanisms in place to develop resiliency against periodic natural disasters. Obviously, completely preventing natural disasters is impossible, but a community can certainly develop plans to keep physical damages to a minimum and reduce interruptions to productivity. Such actions ensure that quality of life is maintained and that the situation returns to normal as fast as possible. Sustainable communities accept disasters as an integral part of the larger environments in which they exist and are willing to assume responsibility for disaster recovery without completely relying on governmental and non-governmental organizations (NGO) assistance.

Many theoretical points of view acknowledge the fact that developing regions of the world, including Latin America and the Caribbean, are unique in their disaster recovery process due to the application of generational labor-intensive and skill-based housing methods. These views agreed that the empowerment of disaster resiliency in such settlements is a key principle of sustainability for post-disaster rebuilding (Aldrich & Smith, 2015; Dudley, 1992; Havlick, 1986; Rawal & Desai, 1994; Miculax & Schram, 1998; Schilderman, 1993). The resiliency empowerment of such communities involve a framework that includes: (a) identifying the nature and magnitude of the disaster (b) determining the scope, size, and scale of the sustainable housing program in order to allocate financial and emotional assistance, and (c) determining the complexity of individual and household needs. In the event of a disaster, both individuals and households tend to miss not only more and more of their unique choices, but also the circumstances under which they made those choices.

Accommodating the broad range of such needs is central to any sustainable housing solution. These needs include:

a. Community or Regional Characteristics—The notion that one size does not fit all is particularly valid in traumatic events like natural disasters. The types, nature, and location of housing need to be tailored to fit the local characteristics of community.
b. Available Housing Options—Disaster housing needs to accommodate a sufficient amount of housing options that reflect the local living pattern or be easily adaptable for local lifestyles without major changes needing to be made.

Once the framework is established, specific guidelines should be followed to affirm that shelter operations are primarily a local responsibility and rely on resources within communities. Disaster housing operations should be conducted primarily on a local level. Apparently, most of the Sustainable housing solutions revolve around means and end levels of the local players. The residents of these communities need to come together to integrate a wide array of services and capabilities in order to meet the needs of disaster victims. It is this strong sense of community and a commitment to others that rallies the local support required to deliver these essential services during a disaster.

Housing operations can appear deceptively simple, but are extraordinary complex. The complexity of demands and the nature of disaster housing options require careful planning to meet human needs. On the surface, providing shelter can appear to be as simple as repairing damaged houses or locating a suitable site and erecting shelters for disaster victims. However, in addition to considering local lifestyle requirements, it is also important to provide shelter for people with special needs, service animals, and household pets. Shelter operations require sophisticated design and planning from a wide range of organizations in order to meet urgent needs quickly. To be effective, housing operations require careful coordination between local officers and NGOs. These parties can anticipate housing requirements based on the following factors: (a) the nature and magnitude of the disaster (b) the complexity of individual and household needs (c) community characteristics, and (d) available housing options.

Shelter operations for catastrophic event enter a new realm of designing, planning, and implementation. Catastrophic events demand exceptional levels of effective and timely planning and execution among community, state, and federal officials. Such events can rapidly exceed local shelter capacity and require assistance from the state and federal governments. It is essential to understand how shelter requirements can expand in catastrophic circumstances.

In conclusion, for emerging, developing, or less developed countries, disaster housing poses a unique set of challenges in terms of socio-cultural, economic, and political sustainability that are taken for granted in more developed regions. However, past events provide evidence that the core of any sustainable disaster-housing program should follow certain criteria. Such criteria can ensure a basic level of post disaster recovery, at least until the next disaster. These criteria include:

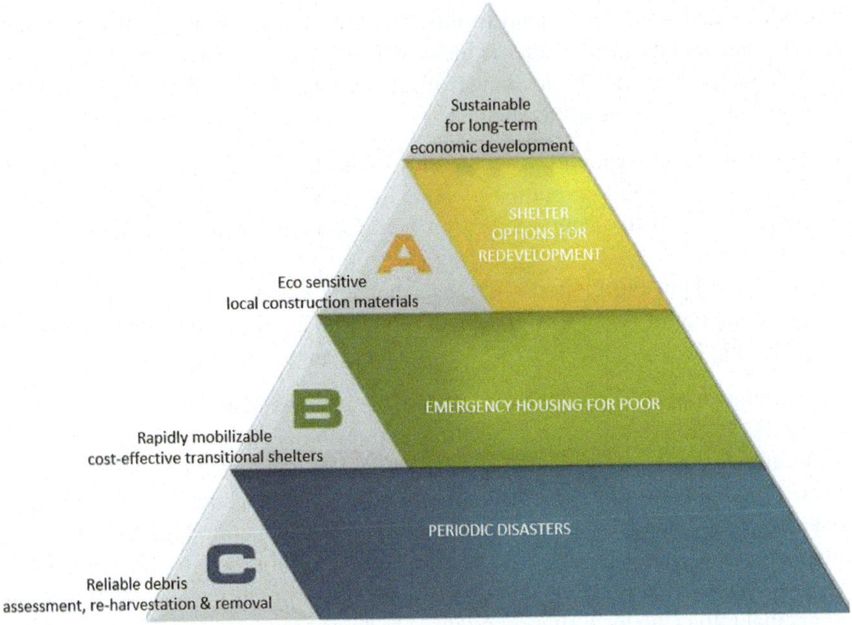

Fig. 2.9 Rebuilding-based mitigation planning for sustainable post disaster recovery

(a) environmentally sensitive construction material (b) sustainability for long-term economic development (c) socio-cultural compatibly to fit the life style of the community, and (d) political vision to mitigate disaster vulnerability. These criteria cross a number of disciplinary boundaries, but guarantee that communities are less vulnerable for future disasters. However, generalizations about the recovery process are not appropriate for extreme natural disaster events. Therefore, each strategy should encompass measures emerging from the event itself. A holistic approach for supporting the economic development of a developing nation would be to train locals in methods of using renewable, environmentally-friendly agricultural byproducts as construction material that can be used to rapidly assemble emergency or disaster housing for displaced victims of a natural disaster (see Fig. 2.9). Furthermore, if the same building materials could be reused, transported, and reassembled to permanent housing, recovery time reduces considerably.

The Purpose

This book seeks to understand the role of sustainability in hurricane recovery of spontaneous settlements in terms of disaster resiliency. The underlying fact is that any phase of recovery from hurricanes presents numerous challenges in terms of rebuilding infrastructures, homes, and businesses. It also opens a limited and time-pressured window of opportunity to redevelop large land parcels cleared during the extreme natural disasters. In addition, such processes provide multiple avenues to rethink the built environment of damaged communities that have been moved away from hazardous areas. Inadequate attention has been paid to the sustainable recovery of evacuees who live in informal settlements of small dwellings under disaster vulnerable conditions. However, the existing body of research reveals these types of settlements display certain disaster resiliency characteristics against natural disasters such as earthquakes and landslides. Building on this, the discourse surrounding the book attempts to ascertain whether these disaster-prone informal settlements provide insight for planning and designing places that are more resilient, physically sound, socially viable, environmentally sustainable, and less vulnerable to future disasters. It particularly focuses on the disaster resilient characteristics of vernacular settlements in terms of sustainable post-disaster recovery.

The discourse further delves into the physical attributes of disaster-prone informal settlements that enable fast and sustainable post-disaster recovery in the reconstruction phase. Therefore, it was hypothesized that the tacit knowledge or traditional "know-how" used to construct dwellings in disaster-prone informal settlements facilitates sustainable recovery from natural disasters. Even though informal settlements' resilience against earthquakes has been well established in the literature for, relatively little research exists regarding the resiliency of vernacular or spontaneous settlements in the United States for hurricanes. An urban-like spontaneous setting located within the territory of the hurricane-prone Bolivar Peninsula of Galveston County was chosen to examine this phenomenon. In the process, cluster patterns and their visual and physical connectivity within settlements were investigated as pertinent factors for disaster resiliency (in terms of sustainable recovery as a suitable case study of the disaster-resilient qualities needed for sustainable recovery and speedy construction with socio-economic efficacy).

Hurricane-Prone Bolivar Peninsula

The existing knowledge on disaster recovery strongly emphasizes the need of saving every possible dollar during tough economic times by adapting strategies leading to the establishment of more resilient settlements with the potential to become more resilient over time. Bolivar Peninsula, a 27-mile long barrier island located few miles northeast of Galveston Island along the Gulf of Mexico, has always been a target destination for hurricanes (see Fig. 2.10). Apart from Hurricane Ike, Bolivar has been subjected to numerous major hurricanes including Alicia in 1983, Jerry in 1989, and Allison in 2001 (see Fig. 2.11). These hurricanes have averaged more than $15 billion in damage to the Galveston Appraisal District. In the early hours of Sept. 13, 2008, Hurricane Ike made landfall as a category two storm on this narrow strip of land. Its infrastructure and the surrounding landscape were critically damaged (see Fig. 2.12).

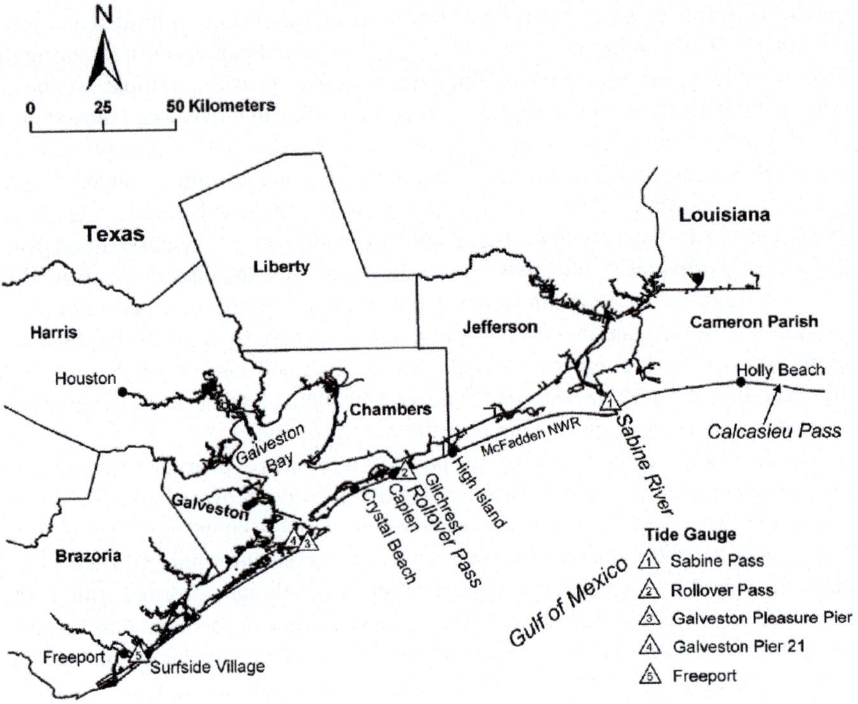

Fig. 2.10 Out of Port Bolivar, Crystal Beach, Caplen, Gilchrist, and High Island, Crystal Beach is the only incorporated community on Bolivar Peninsula (*Source* Kraus & Lin, 2009)

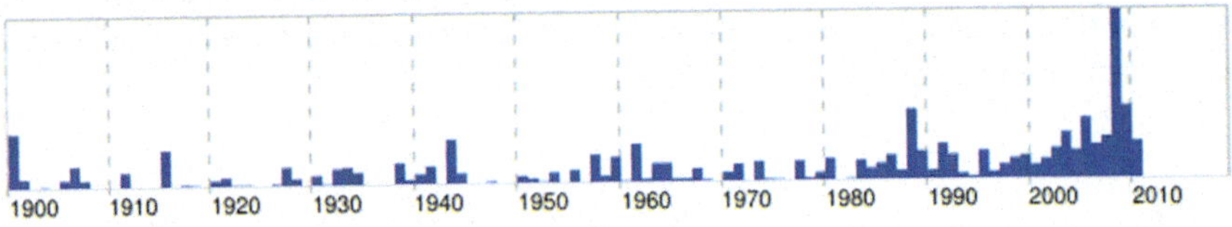

Fig. 2.11 Hurricane history of Bolivar Peninsula from 1900 to 2010 (*Source* Google Timeline Results, 2010)

According to the Institute for Business and Home Safety (2009), the size of the cloud mass and the integrated kinetic energy of Ike were unprecedented and distinctively different from any other hurricane in the Gulf of Mexico observed to date by modern science. The impact of Ike on the Texas Coast, particularly in Bolivar Peninsula, provided ample opportunities for first hand experiences on the performance of building materials, construction methods, and techniques used for reconstruction purposes, as well as for planning and constructing sustainable new settlements. In addition, the magnitude of this event pointed out the need to recognize meticulous and hidden spatial issues such as connectivity, permeability, and integration as potential partakers of sustainable recovery.

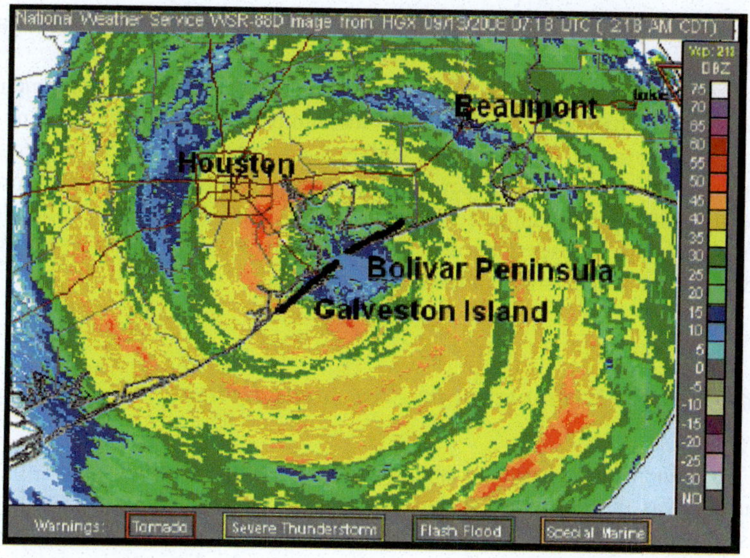

Fig. 2.12 Radar reflectivity image of Hurricane Ike as the eye made landfall on Galveston county (*Source* Marshall, 2009)

Why Bolivar?

Apart from being an unincorporated community, Bolivar has quintessential characteristics of a world in between: the developing and the developed, spontaneous settlements and vacation homes, disaster prone and disaster absorbent, and so on and so forth. Along with its relatively isolated location, Bolivar is a less tainted and affected data source to understand reciprocal dealings between previously stated 'in-between' factors. Hence, it will assist the discourse of rebuilding for resilience for unincorporated communities that are less vulnerable to natural disasters and easily repairable within reasonable reinvestments. On this basis, the proven building traditions of settlements on the Bolivar Peninsula located along the hurricane-prone Gulf of Mexico have been selected. On the other hand, the establishment and growth patterns of the port city-Bolivar peninsula provide glimpses to typical coastal developments of spontaneous settlements along the Gulf of Mexico. The Peninsula was named after Simón Bolívar (1783–1830), a South American hero. Because it was only few miles away from the mainland and the already settled Galveston Island, people went back and forth between the mainland, Galveston, and across Bolivar. The area was particularly important when Antonio López de Santa Anna attempted his conquest of Texas in February 1836 (Griffin, 1931). Before that, in 1816, the peninsula served as a "highway" for the overland slave trade between Galveston and Louisiana. However, the first permanent settler on the peninsula was Samuel D. Parr, who arrived in 1838 and settled in area beginning at Bolivar Point and extending five miles eastward. By 1850, fifteen families lived between Bolivar Point and High

Island, and by 1885 the population of Bolivar Peninsula had grown to 500 (Daniels, 2010).

Even though Bolivar Peninsula shows many unique patterns of development with regard to human habitation, it also shares a common thread with the building traditions of East Texas. In addition, it is widely accepted that East Texas offers many opportunities to investigate the development of regional building habits. Moreover, this region also exhibits a unique architectural style with a large number of buildings. The majority of the houses that survived hurricane Ike belong to a domestic category that demonstrates interesting architectural vocabulary.

Typology and Development of East Texas Houses

East Texas includes 25 counties and stands out among the other regions of Texas because it was the birthplace of an urbane style of building known as the "East Texas House." East Texas is a roughly triangular area demarcated by Louisiana on the eastern border, the Trinity River to the west, and a line passing just north of the cities of Jefferson and Tyler. According to Evans (1952), this region has more geographical commonalities with Louisiana and the Deep South than with the rest of Texas. One of the most important features of this region is its "piney woods," which is a prime source of building materials. This environmentally responsive material was well manipulated in the structural forms of East Texas houses. Its abundance has been a strong influence on the building traditions of East Texas. The warm and damp summers of this region has also had a positive impact on these traditions. Another influence on rural building designs and constructions was the "Texas Norther," a type of cold front known for its rapid southerly movement. East Texas has been successful in creating a house type that is easily ventilated, warm in winter, and cool in summer, as a response to sudden variations in the regional climate (Evans, 1952). Currently, the population groups that control the architectural style of this region are Anglo-Saxon, Protestant, and American-born; however, there were also several Spanish family homes and Spanish mission establishments built in the colonial revival style (Blake, 1939).

The first American settlers who followed Stephen F. Austin to Texas colonized a significant part of East Texas. They came mainly in two groups from what was recognized as the old south. A visible distinction can be identified in the architectural practice between the people who came from the Kentucky and Tennessee hills and groups from further south in Georgia, Mississippi, and Alabama. The first group was larger and typically used a simple wooded country style of building. The second group, mainly plantation owners, preferred grander homes predominantly promoting the Greek revival tradition. According to many scholars, two distinctive types of houses represent the building practices of these early settlers. What was recognized as the "East Texas House" was credited to the Tennessee and Kentucky groups in the north part of East Texas, as shown in Fig. 2.13.

Fig. 2.13 House inspired by Tennessee and Kentucky style: Elliot-Robinson house (C. 1880) Center Road, Nacogdoches County (*Source* Evans, 1952)

The "Louisiana House" type was developed simultaneously among the houses built by the former residents of the Deep South, as shown in Fig. 2.14. These houses were unique in many ways, especially in the adaptation of classical and Greek revival features from prominent-urban landmarks to rural, but contemporary domestic structures. In contrast, the East Texas type followed a different development process similar to houses in Kentucky and Tennessee (Barker, 1969). The skillful artisanship

Fig. 2.14 Louisiana type house: Garrett house (C. 1850), Parker-Dorsey house, San Augustine (*Source* Evans, 1952)

of these constructions can be explained by the scarcity of power machinery during the early settlement of the area. The survival of manual cutting and dressing of lumber in some areas until the present day is a proof this phenomenon. Evans (1952) stated that the "beautiful exterior and interior detail were the result of the carpenter's tools, skill, and taste" (p. 3).

East Texas settlements were characterized by small-widely spaced cities. The road network of early settlements responded to houses rather than houses to the roads. Therefore, in the majority of cases houses faced roads. Despite its many positive geographical characteristics, the region did not attract many new settlers until the oil boom in 1900. According to historians, the remoteness of East Texas and the relatively undisturbed and less sophisticated lifestyle were also important in developing and making the East Texas House outstanding among other types of houses developed at the same time. In addition, the overall climate, including summer, did not demand extra effort in planning and constructing these houses. Comparatively large distances between settlements were characteristic and the late arrival of railroads in the area did not dramatically alter the phase (Reed, 1981).

The East Texas House in its generic form was rural in design and completely built out of wood, including its foundations and chimney (see Figs. 2.15 and 2.16). Quite interestingly, many houses built between 1820 and 1890 still remain in their original primary form. These houses were planned for convenience, and the simple, sound, and

Fig. 2.15 Raised wooden foundations of old Doctor Birdwell House (c. 1840), Mount Enterprise, Rusk County (*Source* Evans, 1952)

Fig. 2.16 Forest Hill (1847), Mrs. Carl Yowell residence near Alto, Cherokee County (top), and Tucker House (c. 1840), Mrs. Granberry Residence, near Chireno, Nacogdoches County (bottom) (*Source* Evans, 1952)

straightforward constructions were reflections of the East Texas lifestyle. They were elevated on wooden pillars, some moderately high, a few low, and accommodated an air space underneath the house. They were well fitted to the land sites, and good ventilation was essential to the design.

A central hallway that was accessible via a front porch divided the typical house layout into two equal sections (shown in Fig. 2.17). Many identified the "hall" as a "dog-run" or "breeze-way." It was common to find a room that opened directly onto the front porch. Single-room houses were another popular design among settlers. Another typical feature to these houses was the orientation of the ridgepole; it often ran parallel to the road and in most cases fairly near it. As seen from the road, the chimney was usually located on the left portion of the house, while the "hall" and "room-across-the-hall" took the right.

The term "double house" represented the structures with all three basic units (house, hall, and room-across-the-hall), which were constructed at one time. However, in some cases, "double house" or "double log house" described the complete structure, even though the elements were constructed at different periods.

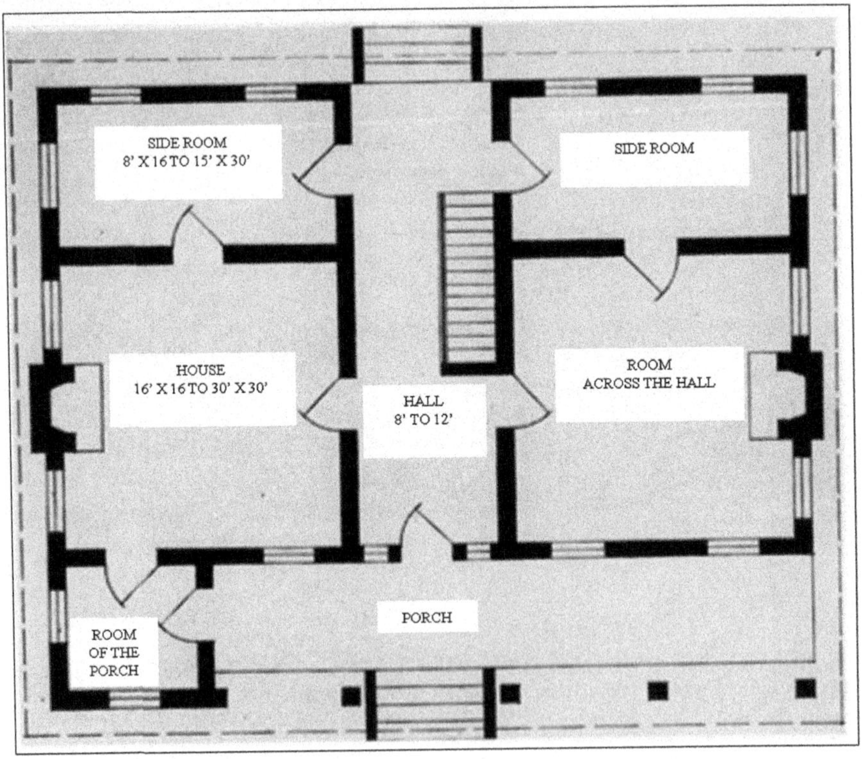

Fig. 2.17 Typical floor plan showing room arrangement and proportions of an East Texas house (*Source* Evans, 1952)

Some two-room houses, and two-room double houses did exist, and some have survived to the present.

The only heating device in these houses was the chimney, which was often constructed as a stick structure filled with dirt and finished with stucco. The inherent fire hazards associated with the construction materials forced builders to locate them four to eight inches away from the house above the firebox. Some of these chimneys are still in use and are strongly representative of regional building habits. Interior chimneys were not considered safe and fireplaces were used for cooking purposes as late as early 1900s.

Some precautions were taken for kitchens by having independent or set-aside kitchens, which were separate structures from the main house. This not only minimized fire risks, but also kept excess heat away from the main house during the summer. The grouping of main house and kitchen was also aesthetically pleasing. After stoves became essential cooking devices, kitchens were incorporated into the houses. However, a few still exist as separate units, even to the present.

Galveston: Twin or Neighbor?

Although strictly urbane, the Galveston settlements are somewhat representative of Bolivar Peninsula residential developments. The areas' first settlers were Native Americans, tribal Karankawas, who came from the area around the Rio Grande River in the nineteenth century, 350 miles away from Galveston Island (Daniels, 2010). This group was extinct by the time Texas became a republic (Cartwright, 1998). They did not erect permanent shelters and moved inland in warmer weathers, moving around when food supplies grew scarce.

Galveston has always been a victim of hurricanes, the most famous of which was the hurricane of 1900. The 1818 hurricane that hit Galveston almost wiped out the entire colony of Spanish pirates led by Jean Laffite. However, they managed to rebuild the settlement using timber from more than half a dozen of ships blown across the bay and torn into pieces. According to McComb (1986), the most substantial building in the town of mostly crude huts was Laffite's two-story frame house. The center of the Galveston settlement was built by Spanish immigrants in the sixteenth century, during the interim government under David G. Burnet; most of these shelters were semi-permanent. Saccarappa, the east end of the island, was occupied named by a few immigrants from Maine. However, Michel B. Menard was the one who adopted the name Galveston and laid out the harbor town in a gridiron fashion for land development purposes after the Texas Revolution in 1835. According to another source it was John G. Groesbeck who brought the gridiron model from Philadelphia to develop the city after Texans won their independence in 1838 (Baker, 2015).

The typical timber framed houses of Galveston fueled the 1885 fire that destroyed nearly a third of the city, including 568 homes in 42 blocks, but the same construction style was instrumental in quickly rebuilding the houses. Building and rebuilding of houses became the biggest industry in the city. Although the builders in the business area were instructed to reconstruct buildings using fireproof materials, people took more freedom doing their homes. The porches of their Victorian style houses were arranged to let gulf winds ventilate living areas, and intricate figures in black, white, red, blue, orange, and pink were used for elaborately decorated house facades. However, they were cautious to use fireproof tin, slate, and galvanized shingles to clad the gable and mansard roofs of their houses (McComb, 1986). Initially, the roads were not paved at all. Later, the city and property owners used "oyster hash" (crushed shells) to pave the roads but this was not very successful. This solution as followed by many others, including wooden blocks in a hot tar setting, bricks, concrete, asphalt, and concrete combination.

The 1900 hurricane completely changed the physiology of Galveston by destroying all segments of the city. The sheer magnitude of the destruction led to the construction of a sea wall along the Gulf edge of the island, as well as using filling sand and dredge materials to raise the grade of the island by eight feet (Mooty, 2005). This was extremely difficult to do under stone structures such as churches and government buildings, though it was less complicated underneath timber-framed buildings. The nineteenth century residential architecture of Galveston is best represented by the late Victorian architectural style. As much as two hundred blocks of residential streets portray the late Victorian inspired residential architecture; this is more than any other major city in Texas. Unlike other Texas cities, new constructions did not typically replace old architecture in Galveston the way normal "progress" does in other cities (Mooty, 2005).

Prominent Galveston architect Nicholas Clayton was commissioned to design grand architectural pieces like churches, academic buildings, and even some residential buildings like the Gresham House for wealthy clientele. Another architect, Alfred Muller, primarily designed frame houses for middle class clientele; he is considered the archetypical Galveston residential style. Although he was not as prolific as Clayton was, Muller's architecture was critical in shaping the bulk of the physical character of Galveston neighborhoods. After Muller's and Clayton's retirement, no new architect came forward to continue, refine, or change their expressive Italianate, Queen Anne, and Richardson architectural styles (Scardino & Turner, 2000).

Undoubtedly, fire and flood, two extreme disasters, have made Galveston's built fabric exotically Victorian. Once the booming oil industry took over the financial momentum of Texas, Galveston did not see further architectural accomplishment as it had when it was one of the main port cities of the North America. Therefore, shortly after the close of the nineteenth century Galveston's new construction and architectural style consisted primarily of imitations of its former glory (McComb, Bixel, & Hurner, 2000).

Conclusions

The official approach to mitigation of disaster recovery among many governments still revolves around scientific knowledge, typically paying more attention to technological issues vs. socio-economic issues. The underlying assumption for such beliefs is the notion that meticulously drafted planning regulations and building codes can adequately address disaster issues. It is not difficult to find many developing regions and underdeveloped regions of developed countries that have adapted such regulations and codes for the exact same reasons. Many federal and local governments tend to borrow planning and building codes from places where they are already in place rather than hiring local experts to come up with their own codes. These prescription ordinances hypothesize that structurally sound construction materials and standardized building methods as the only means to battle extreme forces of nature. On the other hand, such prescriptions do not provide any insight into affordability potentials or cost-effective nature of extremely versatile local building methods of low-income groups that are highly conducive for sustainable post-disaster recovery process. Many difficulties, including a lack of social and economic capacity, are evident in imposed regulations and codes in informal settlements in developing countries, as well as unincorporated settlements in developed countries. Considering that these settlements are occupied by more than half of the global population, one cannot exclude them from disaster mitigation and recovery scenario. On the other hand, many of these regulations and codes demand a higher level of specialized expertise from professionals like architects, engineers, contractors, and building control officers than most of the developing or unincorporated areas. As confirmed by Ruskulis (2002), this often compromises safety in terms of structural soundness of the constructions against natural disasters.

Communities need to be convinced of the positive impacts of planning regulations and building codes and of their necessity in hazard mitigation. Forced implementation of regulations and codes is a proven failure; people find creative ways to circumvent them. In order to increase community acceptance, polices should not alienate local planning, design and execution systems, but should complement them. Therefore, except for policies that are optimal in a given context, overly ambitious rules may not achieve the expected resiliency.

The accumulation of risks is part of the lives of disadvantaged groups; natural hazards are just one part of these risks. It is a fact that such communities must make choices between livelihood options for short-term existence and long-term resiliency against recurrent disasters. In addition, the livelihood and lifestyle processes they participate in, such as coastal fishing and recreation also increase their vulnerability. The notion that natural disasters are one manifestation of vulnerability is more significant than ever. Furthermore, the role of humans in transforming vulnerabilities into mass disasters should seriously take into account any strategies that focus on reducing their impact. However, the majority of conventional approaches to disaster reduction react to symptoms rather than to the underlying causes. Therefore, shifting the view toward mitigation among policy makers and executors is vital. Those who expect to reduce the effects of natural disasters need to understand the relationship between development and disasters as well as cyclical effect of development on accelerated vulnerability.

Past natural disasters have provided ample evidence to support the idea that financing pre-disaster mitigation is more economical as well as less stressful than relief and reconstruction; this is based on the notion that mitigation is not a choice but a necessary part of any developmental schema. Mitigation is more about wisdom of expenditure of funding rather than the monetary value of funding allocation. Top-down methods with heavy administrative structures often promote dependence instead of self-reliance and may even contribute to increasing vulnerability under post-disaster conditions. As they are more case specific, site-based hazard reduction measures often consume fewer resources compared to prepackaged systems with generalized disaster management protocols.

References

Aldrich, D., & Smith, R. (2015). Social capital and resilience. *World humanitarian summit policy brief*. Retrieved May 17, 2019 from http://works.bepress.com/daniel_aldrich/31/.

Alexander, D. E. (2016). The game changes: "Disaster prevention and management" after a quarter of a century. *Disaster Prevention and Management: An International Journal, 25*(1), 2–10. https://doi.org/10.1108/dpm-11-2015-0262.

Al-Nammari, F. A. (2006). *Sustainable disaster recovery of historic buildings, the case of San Francisco after the Loma Prieta earthquake* (Doctoral dissertation). Retrieved June 24, 2011 from http://repository.tamu.edu/bitstream/handle/1969.1/5874/etdtamu-2006A-ARCH-AlNamma.pdf?sequence=1.

Baily, N. M., & Elliot, D. J. (2009). *The U.S. financial and economic crisis: Where does it stand and where do we go from here?* Washington, DC: Brookings Institution Initiative on Business and Public Policy. Retrieved June 12, 2010 from http://www.brookings.edu/papers/2009/0615_economic_crisis_baily_elliott.aspx.

Baker, J. W. (2015). Efficient analytical fragility function fitting using dynamic structural analysis. *Earthquake Spectra, 31*, 579–599. https://doi.org/10.1193/021113EQS025M.

Barker, C. E. (1969). *The life of Stephen F. Austin, founder of Texas, 1793–1836: A chapter in the westward movement of the Anglo-American people*. Austin, TX: University of Texas Press.

Berke, P., & Beatley, T. (1997). *After the hurricane: Linking recovery to sustainable development in the Caribbean*. Baltimore, MD: John Hopkins University Press.

Berz, G., Kron, W., Loster, T., Rauch, E., Schimetschek, J., Schmieder, J., et al. (2001). World map of natural hazards: A global view of the distribution and intensity of significant exposures. *Natural Hazards, 23*(2–3), 443–465. https://doi.org/10.1023/A:1011193724026.

Blake, R. B. (1939). *Historic Nacogdoches* (3rd ed.) Nacogdoches, TX: Nacogdoches Historical Society.

Board on Natural Disasters. (1999). Mitigation emerges as major strategy for reducing losses caused by natural disasters. *Science, 284*(5422), 1943–47.

Buscaglia, L. (1983). *Living, loving, and learning*. New York: Fawcett Columbine.

Cartwright, G. (1998). *Galveston: A history of the island*. Ft. Worth, TX: Texas Christian University Press.

Centre for Research on the Epidemiology of Disasters. (2009). *Graphs of Natural hazard occurrences*. Retrieved April 12, 2010 from http://www.unisdr.org/ppew/iewp/figures.html.

Cutter, S. L., & Emrich, C. (2005). Are natural hazards and disaster losses in the U.S. increasing? *Eos: Transactions American Geophysical Union, 86*(41), 381–396.

Daher, R. (2000). Heritage conservation in Jordan: The myth of equitable and sustainable development. In I. Maffi & R. Daher (Eds.), *Patrimony and heritage conservation in Jordan* (no pp.). Amman, Jordan: Centre d'Etudes et de Recherches sur le Moyen-Orient Contemporain.

Daniels, A. P. (2010). Bolivar Peninsula. *Handbook of Texas Online.* Denton, TX: Texas State Historical Association. Retrieved June 11, 2010 from http://www.tshaonline.org/handbook/online/articles/rrb06.

Dudley, E. (1992). Disaster aid: Equity first. In Y. Aysan & I. Davis (Eds.), *Disasters and the small dwelling. Perspectives for the UN IDNDR* (pp. 146–61). London, UK: James and James.

Ehrenkrantz, E. D. (1989). *Architectural systems: A needs, resources, and design approach.* New York, NY: McGraw-Hill Publishing Company.

Eid, M. S., & El-adaway, I. H. (2016). Sustainable disaster recovery: Multiagent-based model for integrating environmental vulnerability into decision-making processes of the associated stakeholders. *Journal of Urban Planning and Development*, 04016022.

Evans, E. A. P. (1952). The east Texas house. *Journal of the Society of Architectural Historians, 11*(4), 1–7.

Giddings, B. (2000). *Sustaining the cultural heritage of city centre buildings in Northern Europe.* Northumbria University, UK: Sustainable Cities Research Institute. Retrieved February 20, 2006 from http://www.sustainable-cities.org.uk/db_docs/cultureheritagex.pdf.

Giddings, B., Hopwood, B., & O'Brien, G. (2002). Environment, economy, and society: Fitting them together into sustainable development. *Sustainable Development, 10*(4), 187–196. https://doi.org/10.1002/sd.199.

Google Timeline Results. (2010). *Hurricane history of Bolivar Island from 1900 to 2010.* Retrieved February 09, 2010 from http://www.google.com/search?q=Hurricane+history+of+Bolivar+Island+from+1900+to+2010++Source%3A+Google+timeline+results+for+Bolivar+hurricane+history&ie=utf-8&oe=utf-8&aq=t&rls=org.mozilla:en-US:official&client=firefox-a#q=Hurricane+history+of+Bolivar+Island&hl=en&client=firefox-a&hs=y2O&rls=org.mozilla:en-US:official&prmd=ivns&tbs=tl:1&tbo=u&ei=3dAGTtn5FYG6tgfX8M3JCw&sa=X&oi=timeline_result&ct=title&resnum=11&ved=0CGoQ5wIwCg&bav=on.2,or.r_gc.r_pw.&fp=27b93cf21cbb40b9&biw=1600&bih=1007.

Griffin, S. C. (1931). *History of Galveston, Texas.* Galveston, TX: A. H. Cawston.

Haas, J., Kates, R., & Bowden, M. (1978). *Reconstruction following disaster*. Cambridge, MA: The MIT Press.

Hardy, A. L., & Beeton, R. J. S. (2001). Sustainable tourism or maintainable tourism: Managing resources for more than average outcomes. *Journal of Sustainable Tourism, 9*(3), 168–192.

Havlick, S. W. (1986). Third world cities at risk: Building for calamity. *Environment, 28*(9), 6–11, 41–45.

Hillmann, F., & Ziegelmayer, U. (2016). Environmental change and migration in coastal regions: Examples from Ghana and Indonesia. *DIE ERDE—Journal of the Geographical Society of Berlin, 147*(2), 119–138. Retrieved from http://www.die-erde.org/index.php/die-erde/article/view/293.

Hooke, H. W. (1997, May 22). Natural hazards, human impacts, and disaster reduction. *USGCRP Seminar*. Retrieved March 22, 2010 from http://www.usgcrp.gov/usgcrp/seminars/97514DD.html.

Institute for Business and Home Safety. (2009). *Hurricane Ike: Nature's force vs. structural strength.* Retrieved September 19, 2010 from http://www.ibhs.org/content/data/file/hurricane_ike.pdf.

Independent Evaluation Group. (2005). *Facts & figures on natural disasters.* Retrieved July 10, 2011 from http://www.worldbank.org/ieg/naturaldisasters/docs/natural_disasters_fact_sheet.pdf.

Jigyasu, R. (2015). Challenges and opportunities for disaster risk management of cultural heritage against floods. *ICOMOS–Hefte des Deutschen Nationalkomitees, 60*, 22–29.

Kates, R. W., Colten, C. E., Laska, S., & Leatherman, S. P. (2006). Reconstruction of New Orleans after Hurricane Katrina: A research perspective. *Proceedings of the National Academy of Sciences, 103*(40), 14653–14660.

Kesavan, P. C., & Swaminathan, M. S. (2006). Managing extreme natural disasters in coastal areas. *Philosophy Transactions of the Royal Society, 364*(1845), 2191–2216. https://doi.org/10.1098/rsta.2006.1822.

Kraus, C. N., & Lin, L. (2009). Hurricane Ike along the upper Texas coast: An introduction. *Shore & Beach, 77*(2), 3–8.
Kunreuther, H. (2015). Reducing losses from catastrophes: Role of insurance and other policy tools. *Environment: Science and Policy for Sustainable Development, 58*(1), 30–37. https://doi.org/10.1080/00139157.2016.1112166.
Lefevre, M. R. (2000). Pennsylvania's historic preservation plan: Partnerships and public outreach. *Cultural Resource Management, 23*(7), 33–36.
Marshall, T. P. (2006, April). *Hurricane Ivan damage survey.* Poster presented at the 27th Conference on Hurricanes and Tropical Meteorology, Monterey, CA. Retrieved May 2, 2011 from http://ams.confex.com/ams/pdfpapers/106925.pdf.
McComb, D. G. (1986). *Galveston: A history.* Austin, TX: University of Texas Press.
McComb, D. G., Bixel, B. P., & Hurner, H. E. (2000). *Galveston and the 1900 storm: Catastrophe and catalyst.* Austin, TX: University of Texas Press.
Miculax, R. M., & Schramm, D. (1998). Earthquake housing reconstruction and rural development: Joyabaj, Quiché, Guatemala. In M. B. Anderson & P. J. Woodrow (Eds.), *Rising from the ashes: Development strategies in times of disaster* (pp. 225–240). Boulder, CO, & San Francisco, CA: Westview Press.
Mileti, D. S. (1999). *Disasters by design: A reassessment of natural hazards in the United States.* Washington, DC: Joseph Henry Press.
Mitchell, D., Enemark, S., & van der Molen, P. (2015). Climate resilient urban development: Why responsible land governance is important. *Land Use Policy, 48,* 190–198.
Mooty, H. D. (2005). Alfred Muller and Galveston's late nineteenth-century architectural style. *Dissertation Abstracts International,* 43/06, 1857 A, (MAI No. 1431029). Retrieved August 20, 2010 from ProQuest Digital Dissertations database.
Natural Hazards Research and Applications Information Center. (2001). *Holistic disaster recovery: Ideas for building local sustainability after a natural disaster.* Boulder, CO: University of Colorado [Updated February 8, 2005]. Retrieved on April 15, 2003 from http://www.colorado.edu/hazards/holistic_recovery/.
Oliver, S. A. (1998). Successes and failures in post-disaster resettlement. *Disasters, 15*(1), 12–23.
Patton, C. V. (Ed.). (1988). *Spontaneous shelter: International perspectives.* Philadelphia, PA: Temple University Press.
Rawal, V., & Desai, R. (1994). *Repairs and retrofitting: A manual for people of Marathwada.* Ahmedabad, India: Ahmedabad Study Action Group (ASAG).
Reed, S. G. (1981). *A History of the Texas railroads.* Houston, TX: St. Clair, 1941; rpt., New York, NY: Arno.
Rothrock, G. C. (2000). A tale of two communities: Community involvement in preservation planning. *Cultural Resource Management, 23*(7), 23–26.
Ruskulis, O. (2002). Developing processes for improving disaster mitigation of the urban poor. *Basin News, 23*(June), 8–10.
Scardino, B., & Turner, N. (2000). *Clayton's Galveston: The architecture of Nicholas J. Clayton and his contemporaries.* College Station, TX: Texas A&M University Press.
Schilderman, T. (1993). Disasters and development: A case study from Peru. *Journal of International Development, 5*(4), 415–423.
Spatial Hazard Events and Losses Database for the United States. (2011). *Total property and crop losses, 1960–2003.* Retrieved May 21, 2010, from http://www.geophysik.uni-muenchen.de/~malservisi/hazards2006/2005EO410001.pdf.
Toya, H., & Skidmore, M. (2015). Information/communication technology and natural disaster vulnerability. *Economics Letters, 137*, 143–145.
Van der Wink, G., Allen, R. M., Chapin, J., Crooks, M., Fraley, W., Krantz, J., et al. (1998). Why the United States is becoming more vulnerable to natural disasters. *Eos Transactions, 79,* 533–537.
Zickgraf, C., Telenti, V. D., Sara, Ozer, P., & Gemenne, F. (2016). The impact of vulnerability and resilience to environmental changes on mobility patterns in west Africa. Retrieved December 21, 2016, from https://orbi.ulg.ac.be/handle/2268/193650.

Chapter 3
Sustainability-Resiliency Status Quo

> In rough parallel to the principle that local knowledge is not perfect in all respects, another caveat is that neither do all individuals or groups within a given culture or community have perfect control over all local knowledge. (McCorkle, 1989, p. 6)

Building on the notion that environmental sustainability as one of the key contributors to disaster resiliency, this chapter analyses existing research on sustainable recovery following natural disasters. Deviating from general thesis of disaster recovery as a negotiation between the affected and the assistance, this chapter emphasizes redefining post-disaster recovery as a collective choice that communities and external agencies make under pre-disaster conditions before the occurrence of an extreme natural event. Simultaneously, the chapter reviews case studies relating to spatial aspects of disaster-prone settlements that utilize local/traditional/indigenous knowledge or "know-how" as an integral part of achieving sustainability in pre-disaster mitigation, during-disaster management, and post-disaster reconstruction. Scaffolded on this analysis, the chapter further investigates into possible appropriation of community-based knowledge through scientific sources to face changing modes of recurring disasters and the possibilities of introducing such wisdom to localities challenged by infrequent disasters. This will be embarked via real world examples, in order to demonstrate how different aspects of community-based expertise or "know-how" have been developed and adapted to mitigate and withstand the aftermath of natural disasters across different demographic segments. The final leg of the review takes narrow view of post- Hurricane Ike disaster evaluation and the possibility of adapting spatial analysis to reveal the built-in resiliency of Bolivar neighborhoods against hurricanes. Due to previously stated informal cum unincorporated nature of settlements on Bolivar Peninsula, the cases investigated for this portion demonstrates intersected boundaries between spontaneous and vernacular. Also, authenticated through the cited as well as referred sources, the term vernacular has been used interchangeably with informal to discourse methods of construction that use available local resources and practices to address local needs and circumstances. Moreover, this is on a contested ground of scholarship that speculates possible evolution of vernacular over time to accommodate the environmental, cultural, and historical context in which it exists. Noticeably, this evolution has often

C. Subasinghe, *Rebuilding for Resilience: A Barrier Island Case*,
https://doi.org/10.1007/978-3-030-65532-7_3

been marked as crude and underdeveloped, but also has proponents who advocates its importance in rebuilding endeavors.

Sustainable Post Disaster Recovery

Vulnerability is a direct manifestation of the dichotomy between risk and hazards and is best represented through Risk + Hazard = Vulnerability (Maskrey, 1989). The literature related to sustainable recovery after natural disasters focuses primarily on three dominant factors: environment, economy, and community as illustrated in Fig. 3.1. According to Mary (2008), sustainable community development involves several key principles, including "an increasing value of human life and lives of all species, fairness and equality or economic and social justice, decision making that involves participation and partnership, and respect for the ecological constraints on the environment." Although reconstruction or rehabilitation after natural disasters demands a dynamic equilibrium of these factors, uncontested faith on the social component may have an adverse effect on environmental balance. Most importantly, these factors must be considered holistically in any reconstruction effort and should not be compromised for any reason (Berke & Beatley, 1997; Daher, 2000; Giddings, 2000; Hardy & Beeton, 2001; Lefevre, 2000; Mileti, 1999; NHRAIC, 2001; Phillips, 2015; Rothrock, 2000). Because it is multifaceted, sustainability calls for social, environmental, and cultural completeness in many ways. While its social facet calls for social equity, justice, and equilibrium in distribution of resources, its environmental facet calls for sound strategies that avoid environment and resource depletion. At the same time, its cultural facet calls for maintaining cultural resources for future generations and reducing losses due to carelessness, damage, or disasters (Giddings, Hopwood, & O'Brien, 2002).

Environmental Sustainability: The Architectural Call

Why inhabitants of disaster-prone settlements keep building certain type of houses and maintain the same type of spatial organization in their settlements were the driving questions for many studies that have analyzed environmental sustainability and building resiliency. Spontaneous settlements in the context of unincorporated communities, including the ones in the many parts of the United States along the Mexican border, and on Bolivar Peninsula show significantly low socio-economic states. Commenting on such unincorporated settlements, many theoretical bases acknowledge the fact that these regions are unique in their disaster recovery process due to praxis of generational labor-intensive and skill-based building methods. These views capitalize on disaster resiliency defaulted in such settlements as key to achieving sustainability in disaster rebuilding (Dudley, 1992; Havlick, 1986; Miculax & Schram, 1998; Rawal & Desai, 1994; Schilderman, 1993). In addition to

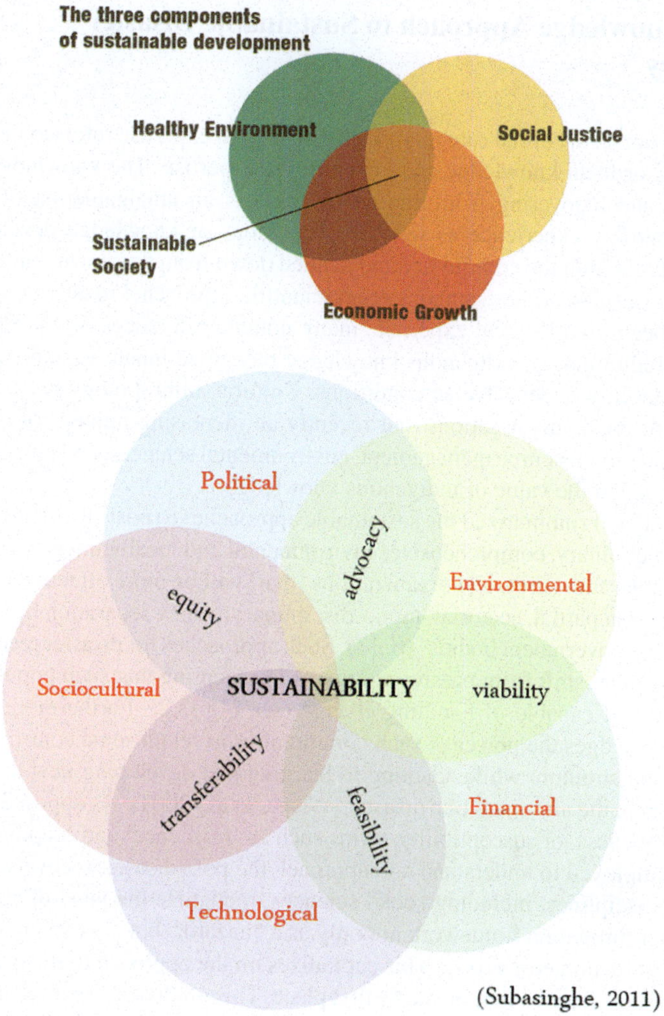

Fig. 3.1 Curtis (2003) sustainable community model against author's sustainability model for disaster resilience (*Source* Curtis, 2003)

this, ample evidence exists to support the argument that any approach that leads to building disaster resilient communities should encompass a holistic understanding of potentially resilient traits inherent in local "ways-of-thinking", "ways-of-doing" and "ways of acting" and therefore built into disaster-prone settlements (Boyle, n.d.; Cuny, 1983; Lee, 2016, Rawal & Desai, 1994; Rogers, 2016).

Local Knowledge Approach to Sustainable Disaster Recovery

In this discourse, the broad category identified as local expertise refers to indigenous, traditional, or local knowledge and generational expertise. The knowledge applied to coping and recovering from natural disasters is an amalgamation of people's own present-day experience as well as generations of knowledge developed and accumulated within the community and handed down from those who have survived disasters in the past. Although mainstream scientific approaches to disaster mitigation have not been kind to local expertise, many notable references on the importance of local, traditional, or indigenous knowledge recognize innate versatility for their easy adaptability under adverse conditions. Cultural anthropology has been at the forefront of these investigations, but recently an increasing number of disciplines including natural resource management, environmental sciences, and human ecology have recognized the value of indigenous knowledge.

Lewis (1982) summarized the sustainable approaches to post-disaster recovery as "a multidisciplinary, comprehensive, environmental and locally integrated approach by indigenous authorities and organizations [that] will be more effective for disaster mitigation than partial, sectorial, mono-disciplinary, policy separation by exogenous agencies and government bodies" (p. 13). Such approaches for disaster recovery have shown a gradual shift from passive recipients to accepting that such populations are to some extent capable of handling change (see Fig. 3.2). Furthermore, this shift also acknowledges the power of such communities to retain some control over their function and structure while adapting to learning and developing new mechanisms for survival in the aftermath of a disaster (Wisner et al., 2004). As opposed to vulnerability, proneness, or susceptibility, terms such as 'resilience' and 'coping capacity' are now employed to understand and approach the post-disaster recovery process.

Many disciplines, including social sciences, recognize the need of approaching the disaster mitigation framework not only as a "hazard," but also as an event that is surrounded by numerous risks. This capitalizes on the approach to the question that asks why people are vulnerable in the first place. This gradual move towards situating

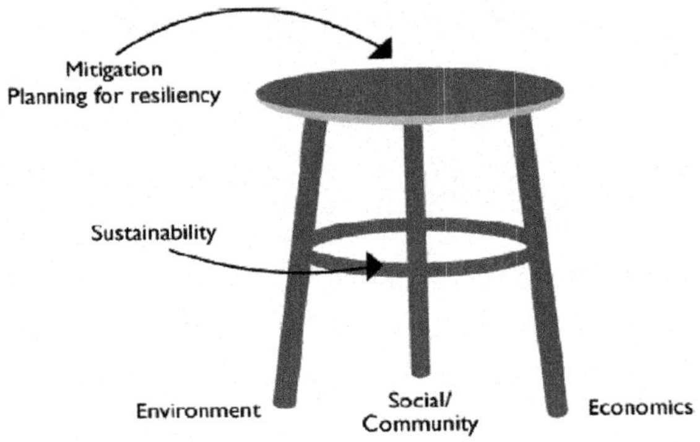

Fig. 3.2 Resiliency values, mitigation, and planning for sustainability (*Source* Godschalk, Kaiser, & Berke, 1988)

disaster risk reduction in the problem area itself has progressed enormously in the last 20 years, especially for small island developing states (Mercera, 2007; Poncelet, 1997; South Pacific Applied Geo-Science Commission, 2004).

Despite a general consensus to use indigenous knowledge to help disaster mitigation, most researchers, decision makers and execution agents have not known how to approach it from the sustainability end of the spectrum. However, indigenous disaster risk reduction systems could, if appropriately modified and adapted and then further complemented by appropriate Western strategies, form the basis of suitable, as well as sustainable, disaster risk reduction strategies. Indigenous systems are typically deep-rooted within mundane activities and at times so subtle that the impact of such practices gets little attention from outsiders, which may be one reason why these practices go unnoticed in the mainstream literature (Johannes, 1978). In particular, the influence of indigenous knowledge embodied in spontaneous settlements and their vulnerability to natural hazards has not been rigorously investigated (Wisner et al., 2004). As noted by Mackinson and Nottestad (1998), the failure to recognize the merits of and employ local knowledge may have affected finding and allocating resources that could have been used to collect applicable information.

It is apparent that the greatest barrier to holistic recovery is the unwillingness and inability to use 'nonscientists' or 'indigenous knowledge' as data (Mercera, 2007). While it is disappointing that such a wealth of information is often ignored or uncounted in its entirety, this disregard may be due to the belief that only data collected in a scientific fashion can satisfy the requirements of statistical analysis. On the other hand, because the knowledge of rural or indigenous groups does not conform to the standard or format expected, decision makers often find it difficult to deal with it (Mackinson & Nottestad, 1998; Schmuck-Widmann, 2001). Many disaster relief schemes and programs treat people as 'clients' in the disaster management processes. During such processes, science and technology do things to and for these 'clients,' treating the affected groups as passive recipients rather acknowledging them as active participants (Weichselgartner & Obersteiner, 2002). Shah (2003) argued that well-known and well-understood techniques must be utilized to make a direct impact on those living with various natural hazards. According to Mercera (2007), the primitive practices of a particular civilization are based on numerous trials and errors that led to specific knowledge of how to deal with disasters, but it may be unintelligible to another civilization. It was only during the late 20th and early 21st centuries that the potential usefulness of indigenous knowledge within development was recognized in the scientific paradigm (CARE & PSO, 2011; Gray & Morant, 2003; Pfeifer, 1996; Sapre, 2000; Singhal, 2000). Recognition of this same potential has come even more recently in emergency management and disaster risk reduction (Cronin, Gaylord, Charley, Alloway, Wallez, & Esau, 2004; Ellemor, 2005; López-Carresi, Fordham, Wisner, Kelman, & Gaillard, 2013; Wisner, 2004). Mackinson and Nottestad (1998) also emphasized the need to acknowledge the generational expertise of rural indigenous groups as a primary source of information by leaving behind the biases of other scholars on their natural resource wealth and their own impact upon this phenomenon.

Another aspect of this scenario is that large-scale, top-down, technologically-driven reconstruction projects that typify post-disaster reconstruction usually engage outside engineers and builders, introduce new and expensive construction technologies, supplant both local knowledge and local labor, and do not necessarily reduce vulnerability overall. Homes are often built without regard for their effects on the lives and livelihoods of the people who will live in them; the nature and extent of participation is frequently limited (Twigg, 2004). In these, and many similar but smaller projects, the focus is on physical structures (houses) rather than on living and livelihoods (homes). Nevertheless, numerous doubts do exist regarding the sustainability of current post-disaster reconstruction strategies. From the perspective of building resiliency or disaster mitigation, continually pouring capital into projects that reconstruct the same type structures repeatedly might be judged as a maladaptation (Klein, Nicholls, & Thomalla, 2003). In addition, recurring intense hurricanes may further threaten the sustainability of the coast by aggravating beach erosion, which constitutes a vital touristic resource (Buzinde, Manuel-Navarrete, Kerstetter, & Redclift, 2010).

Although often neglected, the abundant availability of cheap labor is crucial for rapid recovery following a natural disaster (Subasinghe, 2013). As reported by Dixon (1991), "almost immediately [after Hurricane Gilbert in 1988], at least 5,000 construction laborers descended upon the coast to work for $1.00 [US] an hour from dusk to dawn" (p. 4). However, low wages may reduce the resiliency and adaptation capacities of workers' households; the only coping strategy for these settlements is to keep rebuilding after every hurricane. The adverse consequences can be observed almost immediately. Confidence and trust in viable traditional construction approaches are undermined. Because inhabitants are dissatisfied with the housing that results, and without opportunities to be involved in making safer construction and retrofitting a priority, the resulting structures can have unnecessary vulnerabilities (Twigg, 2004). As neither safety nor sustainability is attained in many of these schemes, the 'myth of speed' that affects most aid agencies must be resisted (Anderson & Woodrow, 1998, p. 49; Schilderman, 2004).

Poverty has often been suggested as breeding fatalism concerning disasters. In reality, when informed building choices are permitted, most people tend to incorporate safety concerns (Maskrey, 1990). People who have homes built for them, without consultation, information, or choice, will naturally adopt a fatalistic view of the product, including fatalistic views on safety. To overcome these challenges, shelter and other buildings that serve communities must be viewed as processes and not just as products (Davis, 1978; UNDRO, 1982).

A different point of view argues that local knowledge often appears to be insufficient when dealing with infrequent disaster situations, as depicted in the case of the Maharashtra earthquake. Twigg and Bhatt (1988) asserted that people are more aware of some natural hazards than others are. However, even if they are aware of the risks, they often have hard choices to make in terms of choosing between immediate gains in, for example, income or food security, and long-term protection against disasters (Maskrey, 1989). In the majority of cases, local expertise is not always adequate to address issues pertaining to infrequent disasters. In addition, a few authors agree that people have developed better resiliency in dealing with some natural hazards than with others, depending on the magnitude and frequency of the hazards and available external support (CARE & PSO, 2011; Maskrey, 1990; Twigg, n.d.). People often have hard choices to make in terms of choosing between immediate gains and long-term resiliency despite their awareness of such risks. Marginalized communities often straggle to choose between income or food security and long-term protection against disasters. Contrary to what many think, the disadvantaged groups do not increase their risks on purpose, but in their battle for survival they often have to accept a certain amount of risk; the alternative risk would be going without food or income (Maskrey, 1989).

Modes of Resiliency

This section of the literature review is based on the recognition that indigenous mitigation and coping mechanisms do exist, as seen in the following examples drawn primarily from Small Island Developing States (SDIS). The main disaster mitigation strategies can be grouped into several general categories for rural indigenous communities. Of these, land use planning, building methods, and environmental resilience deal directly with the spatial aspect of the mitigation spectrum (South Pacific Applied Geo-Science Commission, 2004). Interchangeable use of terms such as sustainability and resiliency in literature explains various sociocultural manifestations of different realities on different geopolitical agendas of reporting and research agents and agencies.

Land Use Planning

Hazard vulnerability often influences both locations of villages and types of housing. In most situations, communities build their settlements on high ground to avoid storm surges and floods, or in areas not prone to landslides. In volcanic locations, people prefer areas that are least vulnerable to lava flow and where prevailing winds will not deposit ash or acid rain on crops (South Pacific Applied Geo-Science Commission, 2004). However, this changed drastically during the nineteenth and twentieth centuries because of high anthropogenic demands such as overexploitation of natural resources and increased populations. Inevitably, such phenomenon pushes people into more hazardous, marginalized areas for habitation. In addition, increased exposure to environmental hazards can also be another outcome of the insensitivity of land planners in terms of their inability to foresee the value of indigenous knowledge (Mercera, 2007).

Spennemann (1998) describes how settlement patterns of South Pacific Marshall Island dwellers are generationally governed by environmental considerations such as wind, wave action, and storm protection. However, this began to change during World

War II when settlements quickly shifted to different strategies than their traditional wisdom. This happened because Japanese and U. S. troops selected areas for their military bases based on security reasons and favorable lagoon conditions for large ships and seaplanes. These military bases had densely populated residential units but unlike the traditional settlements, they were exposed to the windward side of the atolls. This exposed the settlements to both typhoons and high tides, resulting in severe damages.

A massive earthquake in 1993 caused severe damage in western India. Eight years after the resettlements, a number of problems were observed in the rural villages, mainly as a result of relocation (see Fig. 3.3). This happened after agricultural land was acquired from existing villagers, which displaced from their own land; a few of them relocated far away from their agricultural lands. In addition, the spatial plans for design of the new houses were totally different from their traditional way of living. A "city like" development was adopted with a wide street grid and row housing, leaving no space for the activities of specially-skilled groups such as artisans. For these and other reasons, the villagers returned to to their old villages to rebuild on ruins, using their familiar techniques. The most critical concern during this whole process was that none of the "earhquake-resistant" features were adopted in the new reconstructions. Perhaps underestimating the readily available local knowledge made the situation worse than it was and rendered the information dissemination and technology transfer meaningless.

Reporting on the relocation of two villages in Flores, Indonesia, following the 1992 earthquake, Boen and Jigyasu (2005) discussed a critical social issue that was a direct result of the unique combination of the settlement's location and house designs. The reconstructed houses in Wuring, owned by local fishermen, were relocated 200 meters away from the shoreline in order to avoid damage from future tsunamis. The location and housing options were decided primarily by local army officials, who were completely opposed to the traditional pole-style housing of the fishermen. The new houses were constructed in a military barracks style, which was very difference from the traditional housing styles. A fish auction building, and other public facilities were also built. Similar to many well-documented cases, despite the additional facilities, the villagers in Nangahure abandoned the new settlements (Abe & Shaw, 2014).

A similar situation occurred in one of the Babi Island villages when the villagers from two communities were relocated to Nangahale on Flores Island. The villagers from Babi Island were mainly fisher folk. Both communities were from the same village and grew crops on the hillside as their main livelihood. Typically, their dwellings were constructed on poles in order to protect the houses from flooding during high tide. The relocation to Nangahale forced the villagers to change cropping patterns and at the same time to live further away from the sea, which made up part of their livelihood. Religion also played a part in the villagers abandoning the location, due to signification tension between the two groups. Most of the native villagers from Babi Island were Muslims and the local residents of Nangahale were Catholics.

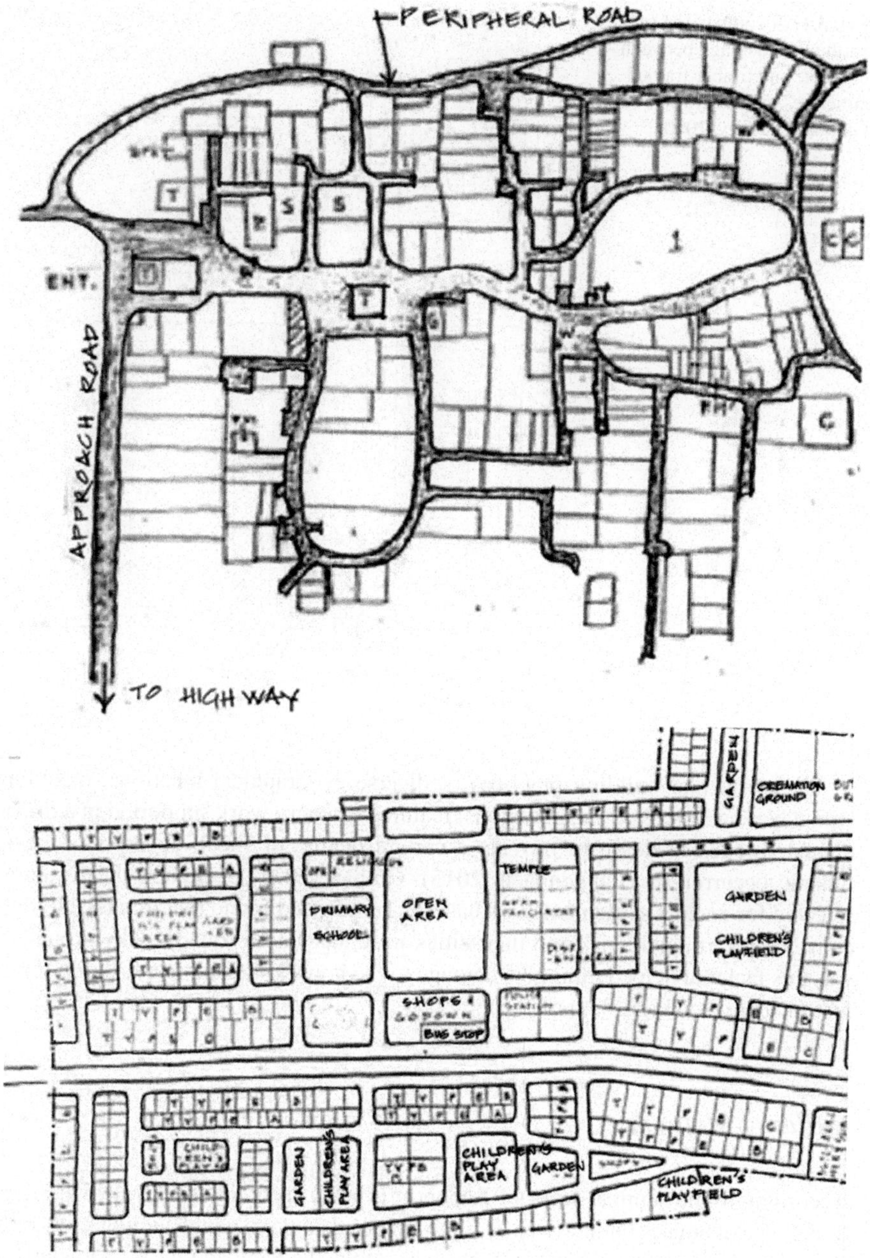

Fig. 3.3 Layouts of earthquake prone traditional (top) and relocated village (bottom) in West India (*Source* Boen & Jigyasu, 2005)

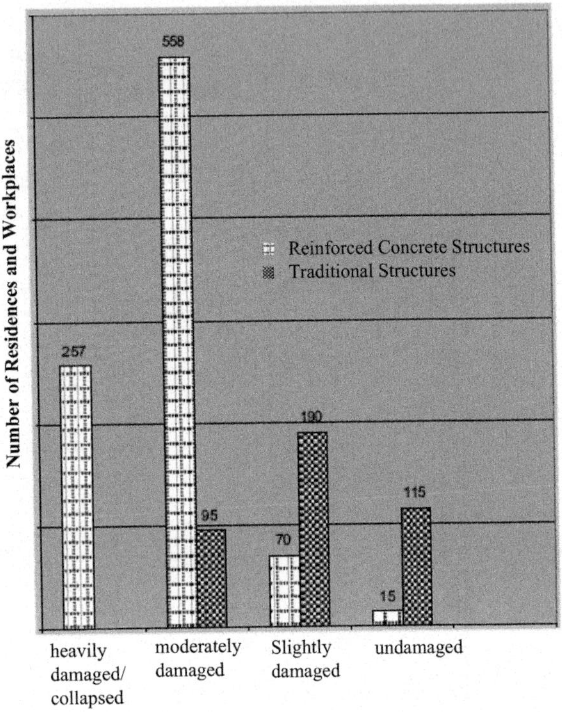

Fig. 3.4 A comparison of earthquake damage between concrete and timber framed structures in Turkey (*Source* Gülhan & Güney, 2000)

The vernacular building practices of diverse geographical locations, including Himalayas and Mediterranean regions, features masonry work supported in a firmly braced wooden skeleton to provide a certain degree of shock absorption during seismic occurrences (Langenbach, 2015). Gülhan (2007) examined the locations selected for ancient settlements in Anatolia, Turkey, and found that that the builders considered earthquake risk and their cities were located on sound, rocky ground. It was this fact that helped minimize damage to the houses situated in the I'zmit urban historic core (see Figs. 3.4 and 3.5).

Building Methods

According to the seminal work of Rapoport (1969), each culture has its own traditional type of house conducive to its socio-physical context, which includes fitness to local environmental conditions. One such example, the Samoan *fale*, is characterized by a high thatched roof that provides cover against sun and rain, but no walls. This allows breezes to flow across the open space, creating well-ventilated interiors. In contrast, the *kanak* in New Caledonia has thick walls, no windows, and a fireplace inside to keep the house warm on cold nights (Iyengar, 2015). In addition, the smoke from the fireplace helps keep out mosquitoes (Dahl, 1989). However, what is common

Fig. 3.5 Timber framed ground floor of an undamaged building in Akcaalan, Turkey (*Source* Gülhan, 2007)

to these traditional dwellings is their character, which is regulated mainly by the occurrence and frequency of tropical cyclones. It has also been observed that indigenous communities in cyclone prone areas developed wind resistant housing where caves were not sheltering option (South Pacific Applied Geo-Science Commission, 2004). One such example is the traditional house of Fiji, the *bure*, which is extremely resistant to strong winds. The *bure* consists of well-concealed strong hardwood posts, a steep angled hipped roof, and secure bindings that holds it down firmly. Throughout the twentieth century, modifications were introduced to these structures, including nails and iron roofing. It was found that such alterations made the *bure* more vulnerable to natural hazards, which in turn increased the indigenous Fijians' vulnerability to recurring cyclones (Campbell, 1984; Janif et al., 2016). Fortunately, some native communities have been able to retain and sometimes further develop their indigenous knowledge (Veitayaki, 2002). One such example is how the indigenous communities in Tikopia on Solomon Island have managed to survive the strikes of severe storms; they strategically changed their survival method by seeking shelter under overhanging rocks on higher ground.

Schmuck-Widmann's (2001) comparative study on indigenous and engineering knowledge along the flood-prone Jamuna River in Bangladesh reported on both existence of flood coping techniques and limited interaction between the engineers working on the river and the Char people living on 'chars' or islands within the river (Islam, Hasan, Chowdhury, Rahaman, & Tusher, 2013). According to the study, the engineers were unwilling to accept the applicability of the indigenous knowledge of the Char people.

In 1993, an earthquake devastated many parts of western India, especially rural settlements characterized by vernacular housing (Boen & Jigyasu, 2005; Langenbach, 2015). This area does not get frequent earthquakes, and a major part of the damage occurred because of the heavy mud roofs found on typical homes in the area. Boen and Jigyasu (2005) noted that the engineers took advantage by advising people to use the external help they received (materials and cash) to construct a small new room of high quality rather than using them for repairs. This may due to the considerable reeducation of the workload but however, as single room was hardly enough to accommodate an extended family, so many of the villagers ended up returning to their former homes, which were unsound and posed greater risks to them. In this case, because the major damage was related to the heavy mud roofs found in the vernacular architecture, all local construction practices were rejected by the "official expert egencies" as unsound. Because existing buildings were considered "unsafe," traditional techniques were replaced by modern technologies during the massive reconstruction program for these relocated villages.

Mitchell (n.d.) discussed another case of local flood mitigation measures in Madhya Pradesh in central India, which took place in 2006. Some of the coping strategies found in the flood-prone houses in the city of Indore included the use of high shelves or raised platforms for storage and placing electric connections up high on the walls. Villagers also constructed platforms in their ceilings to keep valuables, food, and mattresses above floodwaters. In addition to this, rather than screwing or nailing down corrugated iron roofing sheets to the wooden framework of the roof, the villagers weighed them down with stones. This ensured that that the corrugated iron sheets, the most expensive part of their houses could be removed for future use if there was any risk of a house being washed away. Though simple, these adaptive measures significantly reduced the impact of disasters on the villagers' livelihoods, thus making their new starts less stressful economically.

It was reported by Schilderman (1993, 2016) that houses (known as *quinchas*) in Alto Mayo, Peru, were built using local technology and were able to withstand earthquakes better than some of the more recently built structures that used modern technologies (see Fig. 3.6).

The *quincha* houses were simple pole and cane frame structures filled in with earth and sometimes covered with plaster, all locally acquired materials. They were originally introduced by immigrants from drier zones. Post-earthquake observations proved that these quincha structures are relatively flexible during seismic actions. Because the quincha is comparatively cheap and can be constructed with local materials and skills, it is easy to replicate and improve in other post-earthquake settlements. Furthermore, the understanding that relatively cheap housing can be mass produced

Fig. 3.6 Quincha house under construction in Alto Mayo, Peru (*Source* Schilderman, 1993)

using self-labor, local skills, and local materials was essential to the success of the reconstruction project. In addition, local artisans needed only minimum training to incorporate improvements into the design.

Cyclone-prone villages in the Philippines also have certain coping strategies, as reported by Hall (1997). The roofs of the houses of these villagers have vertical and horizontal poles in the framework that are held together by tongue-and-groove joints and pegged together with bamboo wedges. However, it was noted that after many years of weather exposure, the bamboo shrunk and the pegs became loose, which contributed to an overall weakening of the house structure. This means that the villagers are constantly repairing the roofs and hammering in the pegs in order to make them ready for the next cyclone, which is an important and necessary job in the village. The village artisans are an important part of this cyclic event, which in turn supports their livelihood. The level of quality matters to the overall structural integrity, and the better the joints work together, the better the chances of surviving cyclones. However, people in older houses are not without their own survival strategies (Leis, 2016). Although these older houses become a bit rickety, the residents have a simple method to brace their houses: external buttresses. They erect these external supports using long, strong bamboo poles wedged under the eaves and anchored firmly to the ground (see Fig. 3.7).

Flood-prone settlements in many parts of the world provide examples of resiliency by necessity. An article in the Asian Disaster Management News (Haider, 2005) reported on the United Nations Development Program (UNDP) and one of its re-housing programs, stating that flood resistant construction methods of existing houses was instrumental designing new houses after floods in Bangladesh in 2004. The new

Fig. 3.7 Use of bamboo poles to brace structures during storms in Philippines (*Source* Schilderman, 2004)

construction employed existing simple and cost-effective features, which included stabilizing the earthen plinths with cement additive capping. This made them resistant to erosion and the building concrete footing, locally known as *kaatla*, protects the bottom of the bamboo poles by raising them above the plinth level and making structures long lasting. The traditional methods of seasoning were enhanced; the bamboos used in these new dwellings were further treated with local sump oil, kerosene, and bitumen, thus making them more durable (Hamid, 2015).

Many parts of the world, especially developing regions, use beachfront constructions that are light and are typically made of bamboo and thatch. This prevents occupants from being hurt by a collapse of the house due to natural hazards. Therefore, in 2005 the United Nations Educational, Scientific and Cultural Organization (UNESCO) suggested that the use of such traditional construction "rules" should be part of village rehabilitation work. However, even though many key organizations such as UNESCO have emphasized the need to learn from and incorporate indigenous knowledge in restoring villages damaged by natural disasters, in practice very little of this has been done (Méheux, Dominey-Howes, & Lloyd, 2007).

As Gülhan (2007) posits, timber lintel, braces, and nails used in traditional Anatolian houses increased the earthquake resistance of the structures. Similar outcomes were found in research conducted on traditional houses in Japan. Instead of metal ties, screws, nails posts, and beams are connected with mortise and tenon joints in these traditional houses. The *Shinkabe* technique uses bamboo trunks covered with mud plaster as infill. In *Okabe,* an improvement to a traditional technique, the walls are clad with timber boards and plaster, which is applied on metal laths. Another earthquake resistant feature used in these houses is the heavy roofs covered with traditional Japanese tiles or cement-based tiles/plates or with a metal covering (Koçak, Karaca, Pazır, Zengin, & Kalyoncuoğlu, 2013; Takeyama, Hisda, & Ohsaki, 1960).

Malhotra and Ritchie (1980) found that the flexibility of wooden structures was increased when nails were used instead of metal clamps, screws, or joints. This leads one to consider the preference of mortise and tenon joints for architectural elements such as doors and windows in Anatolian houses, and the use of nails for the timber-framed section, as a well-thought-out choice. According to Gülhan (2007), the lath and plaster technique (*bagˇdadi*) incorporated into the timber-framed system further increased structural resistance against lateral forces. In contrast, the use of stone, brick, or mud brick as filling material was found to increase the damage to the buildings. In some cases, it might even have caused the entire collapse of buildings during an earthquake because these heavy materials carry higher self-weights and they are sometimes joined with weak mortar. In future interventions, the use of lighter filling material such as timber lathing could be important. It is not difficult to link such resiliency to earthquakes to deep-seated traditions that are based on many years of experience.

Prior to introduction of mass distribution of petroleum in early 1900, people had to live with what fuels were readily available and devise various methods in constructing their houses so their homes would be compatible with the local climate. The emphasis should be on understanding the passive and low energy characteristics of these settlements in order to learn their spirit to overcome difficulties. The versatile designs of such settlements are adapted to the respective regional climate conditions. Therefore, the importance lies in using high-grade scientific advancements to devise eco-techniques that will mimic these characteristics by (Kimura, 1984).

Environmental Resilience

Because the residents of most rural communities are heavily dependent upon the environment for their livelihoods, it is natural that they develop a wealth of specific knowledge to help them identify signs of impending trouble. Residents of the volcanic islands, Ambae and Vanuatu, are sensitive to environmental signals that hint at potential volcanic eruptions. These signals include pervasive gas smells, the death of trees, unusual active bubbling within the lake surrounding the volcano, rumbling and booming from the crater, and the rapid rotting of taro roots in the ground due to increased ground temperatures (Cronin et al., 2004). This alertness, largely in the form of *kastom* knowledge to oncoming extreme environmental situations has been instrumental in the survival of indigenous populations; it ensures adequate time for the communities to prepare for the impending hazard event (Rungamanee & Cruz, 2005; Walshe & Nunn, 2012).

The Indian Ocean Tsunami of 2004 provided an explosive awareness where scientists were not as effective as certain communities were in warning of the event; a number of indigenous groups survived due to their accumulated knowledge and sensitivity to extreme changes in the environment. Their survival may even have depended on their in-depth knowledge of the dynamics of natural disaster and physical environment transactions. McAdoo, Dengler, Prasetya, and Titov (2006) reported that only seven people were killed on Simeulue Island in Indonesia's Aceh province during the tsunamis of December 26, 2004 and March 28, 2005. This survival rate was strikingly significant considering the absence of any warning system and its close proximity to the nucleus of the earthquake; the northern end of the island was only 40 km south of the earthquake's epicenter. One explanation for the high survival rate was local oral histories, which recounted a similar event in 1907. These traditions advised people to move uphill after prolonged shaking of the ground, which helped save countless lives (McAdoo et al., 2006; Walshe, & Nunn, 2012). Similar incidents of survival attributed to intergenerationally transmitted oral histories and myths have been documented in the Andaman and Nicobar Islands (Bishop, Sanderson, Hansom, & Chaimanee, 2005; Dybas, 2005; Lauer, 2012). This strongly suggests that scientific disaster mitigation studies could benefit from the study of indigenous environmental knowledge and incorporate such wisdom into the risk reduction framework. Even with significant losses to their population and livelihood, many indigenous groups managed to survive drastic environmental changes, as well as develop resiliency over a time and adapt accordingly. In the remote fishing villages of Andhra Pradesh, India, to secure thatch roofs, people throw nets over them weighted down with stones against strong winds during cyclones (Schilderman, 2004).

In a similar case of traditional knowledge being incorporated with scientific knowledge for volcanic hazard management on Ambae Island, Cronin et al. (2004) found that Vanuatu was somewhat successful in handling the aftermath of an eruption. Because the community focused entirely on the hazard itself rather than the underlying components of vulnerability, their success was due to using participatory approaches to produce readily understood hazard maps and community volcanic emergency plans (Walshe & Nunn, 2012). As previously mentioned, the Alto Mayo, Peru case established the fact that most local building knowledge is held by artisans and builders within communities. They understand the various disaster-resistant building technologies better and have ample experience with their performance in disasters. It further emphasizes the importance of identifying such knowledge-bearers and making use of their expertise, however imperfect or little, and adding to it (Schilderman, 2016). The skills base of these communities needs to be retained and sustained in developments of new technologies; this is an important component

Fig. 3.8 Raised platform flood prone house in Bangladesh (*Source* Hufstader, 2008)

of mitigation, and the local artisans and builders are indispensable contributors to that base.

Schilderman (2004, 2016) also found that the flood-prone *kaatla* houses of Bangladesh were relatively cheap and affordable to maintain, even incremental improvements were possible (see Fig. 3.8). The use of local materials and artisans was not only cost effective, but also had value as a demonstration, which was evident because neighbors visited to observe the construction work. Part of the success of these recovery programs was due to the employment of women in the construction, which gave them a sense of ownership. However, Howell (2003) reported that the possibility of employing coastal Bangladesh villagers' traditional early warning signals of cyclones has not been investigated rigorously.

Clear examples of traditional warning systems can be seen in the survival of inhabitants of Tikopia Island, Solomon Islands following a tropical cyclone and the Sea Gypsies in Yan Chiak, Myanmar following a 2004 tsunami (see Fig. 3.9). The Sea Gypsies credited their survival to heeding tales of monster waves created by the spirit of the sea (Anderson-Berry, Iroi, & Rangi, 2003; Dybas, 2005; Hiwasaki, Luna, & Shaw, 2014; Rungamanee & Cruz, 2005; Yates and Anderson-Berry, 2004). Dybas (2005) reported that the spirit warned the gypsies about the earthquake and the quick receding of the sea was the gypsies' cue to run for safe places uphill. In a study of the survival of the gypsies, UNESCO's Regional Advisor for Culture in Asia and the Pacific, Richard Engelhardt, commented on the potential input from local knowledge in disaster management. He said, "The fact that the sea gypsies survived, while many others did not, points to certain lessons to be learned from traditional, indigenous knowledge" (Rungamanee & Cruz, 2005, p. 22). However,

Fig. 3.9 Before and after images of the 2004 tsunami that affected Moken villages (*Source* Rungamanee & Cruz, 2005)

when relocated, the new villages pushed sea gypsies away from beaches into forests; their new housing had poor ventilation, zero visibility, and poor sanitation.

Berkes and Jolly (2001) provided a comprehensive description on successfully developed coping mechanisms and adapted livelihood strategies for a Canadian Inuvialuit people of a small indigenous group living in the western Arctic. In order to manage climatic changes in their community, this group used past knowledge and lived-in expertise gained in this harsh environment (Pearce, Ford, Willox, & Smit, 2015). The conclusions presented in Ellemor's 2005 study on indigenous communities in Australia suggested reconsidering the use of existing emergency management methods. Other researchers have also investigated the potential role of traditional knowledge and the effectiveness of traditional coping strategies in post disaster rebuilding process with a special reference to Asia-Pacific region (Cronin, Petterson, Taylor, & Biliki, 2004; Howell, 2003; Jigyasu, 2002; Lebel, 2013).

The Case of Bolivar Peninsula

Bolivar Peninsula, a 27-mile long barrier island, located in a few miles northeast of Galveston, Texas reveals the opposite of the same coin and thus clarifies "the divided" debate. Although, situated in the developed part of the world with normative higher of quality of life, Bolivar is a quintessential example of universal disaster dynamics of unincorporated communities who share more underdeveloped or 'developing' attributes with their third world cousins along the USA-Mexico border. In the early hours of September 13, 2008, Hurricane Ike made landfall as a category two storm on this narrow strip of land. The island's infrastructure was critically damaged, including residential properties. Afterwards, the Institute for Business and Home Safety (IBHS) reported that the size of Ike's cloud mass and its integrated kinetic energy was unprecedented and distinctively different from any other hurricane in the Gulf of Mexico that modern science has observed to date (IBHS, 2009). The impact of Ike on the Texas Coast, particularly Bolivar Peninsula, provided many opportunities to get fresh insights into the performance of building materials, structural and mechanical systems, and construction techniques pertaining to design, and planning of residential settlements (see Fig. 3.10). As the storm hit a number of highly populated areas and pushed well inland with hurricane force winds, it is not surprising that Hurricane Ike caused huge property losses. In fact, Ike ranks as the third costliest hurricane to make landfall in the United States, behind Hurricane Andrew, which caused $23.8 billion (2008 dollars) in insured losses in 1992, and Hurricane Katrina, which caused $45.3 billion (2008) dollars in insured losses in 2005. Property losses from Ike totaled an estimated $12.5 billion across eight states; at least 115 deaths are linked directly to the storm (United States Geological Survey, 2008). The photos shown in Fig. 3.9 illustrate loss of houses, eroded dune face, and landward sand deposits. Similar to Galveston, because of its geographic location, the damage caused by Ike's storm surge and wave action on Bolivar Island was extensive where hundreds of homes were destroyed (Masterson et al., 2014). The elevation of the first-floor level above the water level and the distance from shore were critical factors in determining whether buildings survived the storm surge and wave action. Buildings with the best chance of survival during this hurricane were at or above six meters above mean sea level.

As per 2015, though limited, the most comprehensive analysis and post-hurricane evaluation of the Bolivar Peninsula come from two agencies and one individual: the American Society of Civil Engineers (ASCE), the United States Corps of Army Engineers, and civil engineer/meteorologist Timothy P. Marshall. Their 2009 report for the Institute for Business and Home Safety (IBHS) focuses particularly on fortified homes that had been built on the island. A team of 14 coastal scientists and engineers from ASCE investigated the upper Texas coast within three weeks of the catastrophe.

The investigation focused on properties in coastal areas in the vicinity of Galveston and Bolivar Peninsula, including piers, marinas, coastal infrastructure, and buildings; the damage to these ranged from minor to complete destruction, depending on the location and elevation. According to the final report, Bolivar Peninsula, which was in the northeast quadrant of the hurricane's path, experienced the most severe damage and three communities were almost completely destroyed. One of the most important conclusions from the analysis was that elevation was a key determinant for building survival as well as a major factor for the stability and effectiveness of shore protection. The report also concluded that with some exceptions, much of the public infrastructure was still in usable condition following the hurricane (Ewing et al., 2009).

Another study by the U.S. Army Corps of Engineers emphasized that "out of the box" strategies similar to the ones taken after the 1900 storm are needed to develop a sustainable protection plan for the coast and future development in the coastal areas (Tirpak, 2009). The report focused on redevelopment and reconstruction, and suggested moving forward with measures that reduce the amount of storm damage to the coastal communities. These measures include using reinforced construction methods and materials, improving the overall design of residential areas, and building economical and environmentally sustainable structural coastal barriers. Tirpak also emphasized that all the report's recommendations should be carefully evaluated and

Fig. 3.10 Before and after Ike images of one of the studied settlements on Bolivar Peninsula (*Source* Doran et al., 2009)

that a coordinated effort among local, state, and federal agencies is needed to prepare the region for the next major hurricane.

Reporting on observations made during Hurricane Ike, Marshall (2009) explained that items most susceptible to wind damage (carports, porch overhangs, and roof coverings) were affected first. However, while laminated asphalt shingles and metal roofing outperformed three-tab shingles, exterior insulation finish systems (EIFS) and vinyl siding were also damaged. As per Marshall's report, wood-framed residences elevated on timber pilings are the most common type of buildings on Bolivar Peninsula and west Galveston Island. In most cases, elevated floor platforms vary from two to four meters above grade, or four to six meters above mean sea level. Many houses were victims of both storm surge and wave action and proximity to shore and height of elevation played major roles in their vulnerability. Those properties at the lowest elevations and closest to shore were most likely damaged or destroyed by storm surge and wave action. Marshall (2009) also identified increasing degrees of damage to building foundations. First, wave action caused removal of sand from around the bases of pilings and from beneath concrete slabs on grade. This caused partially or completely suspended concrete slabs that were exposed between the pilings. Due to the increased wave action, these concrete slabs were broken into pieces. Ultimately, the pilings failed because of continuous removal of sand from their bases. Pilings with shallow depth were quickly undermined by the wave action.

Irrespective of the wave zone, damage to elevated floors increased when the waves reached floor level. First, the wave action managed to rotate or remove blocking between the floor joists. This was followed by uplifting and dismantling of the floor covering as fasteners were incrementally withdrawn from the covering. In some cases, higher waves not only uprooted and removed floor joists, but also pushed them towards the landward sides of the houses. Laterally stacked floor joists along the landward sides of the houses were a common scene during post-Ike inspections. Once the critical floor structures became unsettled, it was inevitable that the homes collapsed or were left perched precariously on top of the pilings. According to the United States Geological Survey temporary monitoring network, the sensor in the middle of the Bolivar Peninsula recorded a maximum storm surge of approximately 4.7 meters above mean sea level with peak wave heights up to 5.9 meters. This meant that the elevated floors of coastal buildings had to be at least six meters above mean sea level to escape the storm surge and wave action (Doran, Plant, Stockdon, Sallenger, & Serafin, 2009).

Reports by the Institute of Business and Home Safety (2009) and Marshall (2009) indicated that a number of homes survived Hurricane Ike on Bolivar Peninsula and west Galveston Island. Noticeably, these homes were located further from the scour zone and had elevated floors more than six meters above mean sea level. In addition, these structures were solidly built with hurricane straps at critical structural points, sound wall cladding, and hip type roofs covered with metal or laminated type asphalt shingles. Many of them also had window and door shutters. Both the IBHS and Marshall reported that wind damage to residential buildings was seen mainly on cladding items such as roof shingles, vinyl or hardboard siding, and unprotected windows. In a few houses, wind action took portions of roof decks away and pushed gable ends either inward or outward. However, because the roof structures usually were strapped to the wall top plates, few roof structures blew off. Some, metal buildings did lose overhead doors that led to loss of metal cladding. Generally, windstorms caused damage to roof coverings and siding, with occasional partial or complete removal of the roof structure.

The main body of literature assessing the Galveston-Bolivar hurricane damage agreed that damage to buildings caused by storm surge and wave action had different characteristics than damage caused by wind. Even though the buildings damaged by storm surge and wave action were dismantled from below, buildings damaged by wind had the greatest damage at roof level. This analysis indicated that magnitudes of forces differ greatly between wind and wave action, but in both cases, building deficiencies were exploited by the storm. These deficiencies included inadequate pile embedment and poor attachment of walls to floors and roofs to walls.

The Institute for Business and Home Safety (2009) constructed a few experimental hurricane resistant houses called "fortified homes" along the Bolivar coastline. They reported minimal or no damage to 10 of their 13 fortified homes located on Bolivar Peninsula. Significantly, these homes were elevated on steel-reinforced concrete columns with the first-floor level placed well above the level of the storm surge and wave action. However, three out of 13 fortified homes received damage from non-fortified neighboring homes that were pushed off their foundations during the storm surge.

According to state data from 2007 prior to Hurricane Ike, Bolivar Peninsula accommodated a combination of 5,425 housing units, including single-family homes, mobile homes, and apartment buildings. Bolivar's housing density is approximately 120 housing units per square mile, and the majority of these units were built between 1960 and 1989. These numbers accounted for 2,091 occupied units, of which 1,511 were occupied by owners and 290 by renters. The rest of the homes were used for seasonal living, mostly by vacationers. However, there was no accurate record for the number of housing units currently habitable on the peninsula. The majority of the houses constructed near the fortified homes were built between 1960 and 2005. Regarding the homes located on the seaward side of Highway 87, the 2009 IBHS report noted that the elevations above mean sea level ranged from 13 feet for those built in the 1960s to 19 feet for those constructed after 1996.

With one exception, all of these houses were destroyed by Hurricane Ike. Highly publicized as the "Last House Standing," the home in the Gilchrist area was the only traditionally built house that withstood the hurricane. It was located in close proximity to Rollover Pass. The house was structurally supported by wooden columns 14 feet about ground level, or 22 feet above the sea level. Some houses built near the coast and located further to the west, where the Bolivar Peninsula widens and the surge was likely a few feet lower, were not completely destroyed. Unlike the coastal houses, many of these houses located further inland were flooded but remained standing. There was no visible debris line in the Gilchrist area, which indicated that the surge completely washed across the neck of the peninsula and that most of the structures probably ended up in the bay.

According to the combined analysis of many studies including Doran (2009), IBHS (2009), Kraus & Lin, (2009), and Marshall (2009), the hurricane caused several types of damages to the houses. Initially, the seaward wall of the first floors failed because of waves attacking the houses slightly above the floor beams, and the bottom part of the walls was pushed partly into the interior of the homes. Both seaward and landward sides of the homes were missing exterior wall cement-board panels. In some cases, this was due to wall siding being fastened to the wood structural panels of the walls rather than to wall framing members. The top of the roof opened up because of the absence of straps over all mating pairs of rafters or because installed collar ties were not installed between every pair of rafters at the ridge of the roof. The water damage to the interiors was significant and flooding left debris inside the house. Concrete slabs were severely damaged at ground level in many homes, which meant that restoration would not be possible without demolishing and removing the slab in order to fill and compact soil properly under the slab.

The issue of damaged concrete slabs has been previously noted by many engineers, who are openly opposed to large concrete slabs being poured under elevated coastal homes. It is critical to avoid undermining the slab, which can cause a house to collapse from wave action. The "Last House Standing," which did not have a large concrete slab, is a good example of this. In addition, the loads that these large slabs can impose on a structure may also cause a significant risk to the structure itself. Instead of one monolithic ground slab under a home, Marshall (2009) recommended using large concrete segments that are essentially giant pavers. The idea is that these pavers can be moved by the surge, and then replaced after soil or sand under the structure is replaced and compacted.

It is possible to make some general conclusions concerning the performance of pile foundations of the partially damaged second and third rows of houses further away from the coast and farther west along the Bolivar Peninsula. Marshall (2009) observed that the houses located closer to the Gulf that were built on pilings only up to the bottom of the first floor tended to be completely wiped out by the surge. However, damage was significantly less where houses were located far enough inland that wave action and surge height dropped below the bottom of the first elevated floor. The upper floors of a number of houses along the coast that had pilings that continued through the height of the first floor remained intact, but in most cases the bottom floor and parts of the walls on the bottom floor were completely gone.

Building Biopsies

The major cause of damage for each structural part of the houses as distilled from the literature is shown in Table 3.1. The following discussion provides detailed descriptions for each built element from selected cases depicted in the literature review.

Table 3.1 Summary table for different structural components and their causes for the damage

	Cause of Damage	Storm	Surge	Wave	Debris attack	Unsound construction
Item	Description					
Walls	Piles on ground		√	√	√	
	Patios			√	√	
	Sliding			√		√
	Doors/windows	√	√			
Roof	Ridge/hip/eave shingles	√				√
	Underlayments	√				√
	Flashings	√				√
	Structural members	√			√	√
	Sheathing	√				√
	Overhangs	√			√	√

Elevation

The IHBS (2009) reported that even homes with 19-foot elevation on the coastal side of Highway 87 barely survived the storm surge generated by Hurricane Ike. Furthermore, if the surge plus the waves topped out at 20 t to 21 feet, a few feet above Base Flood Elevation (BFE), the storm surge and wave action could have completely destroyed all the homes, including the ones that were built prior to 1996. Because extreme surge events are becoming more common, the IBHS further emphasized that there is no substitute for enhanced elevation in cases of extreme surge activity.

Front Porches

The surge either broke or lifted the posts of the porches of many houses out of the ground. Kraus and Lin (2009) noted that in many cases, the broken or removed posts acted as sizeable battering rams and were capable of causing damage to surrounding properties. In addition, when the porch supports or the porches themselves collapsed, they also caused or heightened the damage done to the home itself.

Siding

It was noted that the sidings of "Last House standing" were stripped from some of the upper level walls. In addition to having caused severe damages to roofs, the observed wind damage typically involved damage to hardboard siding. Some buildings with exterior insulation finish systems (EIFS) and vinyl siding also experienced a range of damage from minor to major (Marshall, 2009).

Ridge and Hip Shingles

With one exception, the IBHS (2009) reported that more than 150 houses in close proximity to the western end of the Bolivar peninsula did not have exposed roof decking along the ridge and hips. Subsequently, this allowed rain into the interior, which worsened when the roof decks were exposed. Unfortunately, there was no solid data available on regarding the presence of true hip and ridge shingles. However, many sources stated that most roofs were capped with regular shingles cut to size instead of true hip and ridge shingles (Marshall, 2009). According to the same sources, shingles covering ridges performed better than shingles covering hips. Damages to drip edges and shingles of roof overhangs were shown in photographs; several of these photos provided evidence for damages to drip edges and adjacent eave shingles.

Secondary Water Protection and Underlayments

The underlayments of many roofs survived where shingles failed. However, there were cases reported where the underlayment was also gone and visible water intrusion was significantly present (IHBS, 2009). Some photographic evidence showed the presence of tightly spaced cap nails on the underlayment that survived the wind action.

Flashing

Absence of proper flashing played a significant role in water intrusion and roof leaks. In one verified case, where the roof cover and underlayment were lost, there was clear evidence of badly installed flashings in a number of important areas (Kraus & Lin, 2009). In one such situation, the roof of a particular house included a change in slope about half way down the main roof surface. The lack of flashings over the joint between the primary and secondary slopes led to water collection that caused water to leak through the gap between the sheathing.

Roof Structure

The roof structure of the "Last House Standing" showed a separation at the ridge; this may have occurred because the rafters on opposite sides of the ridge beam were not connected properly. It was estimated that the wind speeds from Hurricane Ike reached 115 mph on Bolivar Peninsula and the wind loads on the roof of this house were estimated as 30% lower than the design loads. However, there was not enough evidence to prove that wind was the only cause for separation on the roof. Interactive forces from both surge and wave action on the house may have contributed to failure of the roof structure.

Roof Sheathing Attachment and Gable End Overhangs

Loss of roof sheathing at gable ends was reported in many homes located toward the western end of Bolivar Peninsula. At the time of damage to roof, the three-second wind gust speed was likely less than 110 mph in this particular area. At the overhang ends of gable failed in many houses, especially roof sheathing was loss at gable ends. In most of these situations, outriggers supporting the sheathing had been notched, and in some instances, the underlying rafters or trusses were likely notched.

Spatial Analysis and Sustainability

Sustainable disaster recovery also requires constituency empowerment so that citizens can advocate for themselves in the face of future disasters. The idea behind this assertion is leaving the recovery in the hands of disaster-stricken communities by enabling systems that are grounded on the site itself, leading to rapid as well as sustainable recovery (Kincheloe & McLaren, 2000; Smith & Wenger, 2006). An objective understanding of disaster-vulnerable settlements' physical fabric enables the process of addressing development issues through sustainable disaster recovery. This goes beyond restoring communities to their previous conditions. It entails the reshaping of the environment, i.e. improving pre-disaster conditions and enacting meaningful changes in communities including emergency management outreach and reaching ou to groups that are typically not being included during the recovery planning (Masterson et al., 2014; Smith & Wenger, 2006).

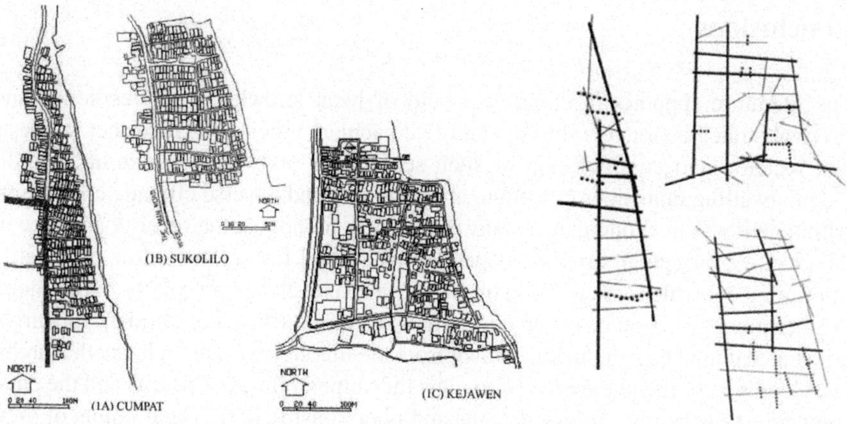

Fig. 3.11 Three fishing settlements (left) and their axial representation (right) (*Source* Darjosanjoto, 2007)

Building and settlement forms are embodied in social-cultural norms of societies (Hillier & Hanson, 1984; Hillier, Hanson, & Peponis, 1987). Analyzing spatial qualities of artifacts including human settlements can reveal the social rules that regulate the interface among people. The analytical procedure is based on graphic representation, nodes, and links, of traditional architectural floor plans or settlement layouts and the qualification of graphs using mathematical formulae. The axial maps generated to investigate the relationship between patterns of physical movement (circulation) within a settlement and airflow within the open-space layout.

Spatial analysis has proven valuable in assessing the vulnerability of settlements from disasters such as earthquakes and floods (Aghataher, Delavar, Nami, & Samnay, 2008). Darjosanjoto (2007) analyzed three fishing settlements in Surabaya, Indonesia, and found that patterns of physical movement (circulation) within a settlement were clearly related to airflow within the open-space layout shown in Fig. 3.11.

The underlying principle that informed this study was the extent to which airflow distribution in a settlement is dependent on the spatial configuration of the settlement. Darjosanjoto concluded that the closer the spaces were to the 'root' of the incoming wind, the shallower the configuration. The opposite was true when only single access was available to many spaces, and the spaces themselves were at some distance from the root (i.e. the area outside of the settlement). However, even though syntactical analysis has been extensively exploited in many other fields of research, this study is one of only a few attempts to link spatial analysis to disaster research.

Conclusions

The literature supports the existence both of local knowledge for resiliency and survival strategies for rebuilding among vernacular settlements. It further suggests that location and configuration of such settlements and spatial make-up of individual dwelling units does constitute ability to respond adverse climatic conditions. Although it was not conclusive, many researchers found that the severity and type of damage to a particular structure depended on several factors, including the quality of original construction, elevation above mean sea level, and proximity to the shore. Furthermore, elevation has been depicted as the most critical determinant for surge-associated damage. Conclusions drawn from the literature review indicate that architects, engineers, and planners tend to view the vulnerabilities of people and the environment as the result of weak designs and poor constructions. Such points of view take on the technical task of fixing the offending buildings and community design. The apparent hypothesis behind this argument involves preventing future disasters by adapting expensive state-of-the-art designs, robust materials, and cutting-edge construction details. Furthermore, building physicists have improved the accuracy and precision of estimates of ground shaking for microzonation and storm surge heights to the nearest centimeter. Increasingly detailed, performance-based building codes have been produced because pioneering advocates from these professions have pursued their adoption as public policy in many countries. Most of the efforts have focused on changing the structures while less attention has been paid to effecting needed change within specific social, political, cultural, and economic environments. The consequence is that the people who are the intended beneficiaries of these advances in both technical knowledge and policies have sometimes become steadily more vulnerable. This suggests the necessity for a community-based approach to reconstruction for disaster risk reduction. It is understood that some examples of community-based mitigation are derived as lessons; such awareness comes from learning from the past, building relations with communities, encouraging participation, involving local builders and artisans, building local capacity, documenting and sharing accumulated knowledge, and most importantly, influencing formal education.

References

Abe, M., & Shaw, R. (2014). Community resilience after Chuetsu earthquake in 2004: Extinction or relocation? In *Community practices for disaster risk reduction in Japan* (pp. 191–208). Springer Japan.

Aghataher, R. M., Delavar, M. R., Nami, M. H., & Samnay, N. (2008). A fuzzy-AHP decision support system for evaluation of cities' vulnerability against earthquakes. *World Applied Sciences Journal, 3*(1), 66–67.

Anderson, M. B., & Woodrow, P. J. (1998). *Rising from the ashes: Development strategies in times of disaster*. London, UK: Intermediate Technology Publications.

Anderson-Berry, L., Iroi, C., & Rangi, A. (2003). *The environmental and societal impacts of Cyclone Zoe and the effectiveness of the tropical cyclone warning systems in Tikopia and Anuta.*

Townsville, Australia: James Cook University Centre for Disaster Studies. Retrieved July 1, 2011 from http://www.pacificdisaster.net/pdnadmin/data/documents/5606.html.

Berke, P., & Beatley, T. (1997). *After the hurricane: Linking recovery to sustainable development in the Caribbean.* Baltimore, MD: John Hopkins University Press.

Berkes, F., & Jolly, D. (2001). Adapting to climate change: Social-ecological resilience in a Canadian western arctic community. *Conservation Ecology, 5*(2), art. 18. Retrieved date http://www.ecologyandsociety.org/vol5/iss2/art18/.

Bishop, P., Sanderson, D., Hansom, J., & Chaimanee, N. (2005). Age-dating of tsunami deposits: Lessons from the 26 December 2004 tsunami in Thailand. *The Geographical Journal, 171*(4), 379–384.

Boen, T., & Jigyasu, R. (2005). Cultural considerations for post-disaster recovery: Challenges for post-tsunami. *Asian Disaster Management News, 11*(2), 10–11. Retrieved June 13, 2011 from http://www.adpc.net/irc06/newsletter/2005/4-6/02.pdf.

Boyle, C. (n.d). *Disaster resistant housing, a handbook of guidelines.* Canberra, Australia: AODRO.

Buzinde, C. N., Manuel-Navarrete, D., Kerstetter, D., & Redclift, M. (2010). Representations and adaptation to climate change. *Annals of Tourism Research, 37*(3), 581–603.

Campbell, J. R. (1984). *Dealing with disaster: Hurricane response in Fiji.* Honolulu, HI: Pacific Islands Development Program.

CARE & PSO. (2011). *Harmonized perspectives: Study for the harmonization of knowledge and experience from indigenous/local, technical and scientific spheres for disaster risk reduction (DRR) and climate change (CC) programming.* http://cdkn.org/wpcontent/uploads/2011/09/Harmonized-Perspectives.pdf. Accessed December 29, 2016.

Cronin, S. J., Gaylord, D. R., Charley, D., Alloway, B. V., Wallez, S., & Esau, J. W. (2004). Participatory methods of incorporating scientific with traditional knowledge for volcanic hazard management on Ambae Island, Vanuatu. *Bulletin of volcanology, 66,* 652–668. https://doi.org/10.1007/s00445-004-0347-9.

Cronin, S. J., Petterson, M. J., Taylor, M. W., & Biliki, R. (2004). Maximizing multi-stakeholder participation in government and community volcanic hazard management programs: A case study from Savo, Solomon Islands. *Natural Hazards, 33*(1), 105–136.

Cuny, F. C. (1983). *Disasters and development.* New York, NY and Oxford, UK: Oxford University Press.

Curtis, F. (2003). Eco-localism and sustainability. *Ecological Economics, 46*(1), 83–102.

Daher, R. (2000). Heritage conservation in Jordan: The myth of equitable and sustainable development. In I. Maffi & R. Daher (Eds.), *Patrimony and heritage conservation in Jordan* (no pp.). Amman, Jordan: Centre d'Etudes et de Recherches sur le Moyen-Orient Contemporain.

Dahl, A. L. (1989). Traditional environmental knowledge and resource management in New Caledonia. In R. E. Johannes (Ed.), *Traditional ecological knowledge: A collection of essays* (pp. 45–53). Cambridge, UK: World Conservation Union.

Darjosanjoto, E. T. S. (2007). Permeability maps of residential settlements within the coastal area of Surabaya, Indonesia. In *Proceedings of the 6th International Space Syntax Symposium.* Istanbul, Turkey.

Davis, I. (1978). *Shelter after disaster.* Oxford, UK: Oxford Polytechnic Press.

Dixon, C. (1991). Yucatan after the wind: Human and environmental impact of Hurricane Gilbert in the central and eastern Yucatan Peninsula. *GeoJournal, 23*(4), 337–345.

Doran, K. S., Plant, N. G., Stockdon, H. F., Sallenger, A. H., & Serafin, K. A. (2009). *Hurricane Ike: Observations and analysis of coastal change.* U.S. Geological Survey Open-File Report 2009-1061. Retrieved January 11, 2011 from http://pubs.usgs.gov/of/2009/1061/pdf/ofr2009-1061.pdf.

Dudley, E. (1992). Disaster aid: Equity first. In Y. Aysan & I. Davis (Eds.), *Disasters and the small dwelling. Perspectives for the UN IDNDR* (pp. 146–161). London, UK: James and James.

Dybas, C. L. (2005). Native lore tells tale. *Oceanography, 18*(4), 12.

Ellemor, H. (2005). Reconsidering emergency management and indigenous communities in Australia. *Environmental Hazards, 6*(1), 1–7.

Ewing, L. C., Stauble, D. K., Work, P. A., Edge, B. L., Rogers, S. M., Loeffler, M. U., … Gregory, G. H. (2009). Field investigation of Hurricane Ike impacts to the upper Texas coast. *Shore and Beach, 77*(2), 9–23.

Giddings, B. (2000). *Sustaining the cultural heritage of city centre buildings in Northern Europe*. Northumbria University, UK: Sustainable Cities Research Institute. Retrieved February 20, 2006 from http://www.sustainable-cities.org.uk/db_docs/cultureheritagex.pdf.

Giddings, B., Hopwood, B., & O'Brien, G. (2002). Environment, economy, and society: Fitting them together into sustainable development. *Sustainable Development, 10*(4), 187–196. https://doi.org/10.1002/sd.199.

Godschalk, D., Kaiser, E., & Berke, P. (1998). Integrating hazard mitigation and land use planning. In R. Burby (Ed.), *Cooperating with nature: Confronting natural hazards with land-use planning for sustainable communities* (pp. 85–118). Washington, DC: Joseph Henry Press.

Gray, L. C., & Morant, P. (2003). Reconciling indigenous knowledge with scientific assessment of soil fertility changes in Southwestern Burkina Faso. *Geoderma, 111*, 425–437.

Gülhan, N. S. (2007). Observations on earthquake resistance of traditional timber-framed houses in Turkey. *Building and Environment, 42*(850), 840–851.

Gülhan, D., & Güney, I. O. (2000, November 16–18). The behavior of traditional building systems against earthquake and its comparison to reinforced concrete frame systems: Experiences of Marmara earthquake damage assessment studies in Kocaeli and Sakarya. In *Proceedings of Earthquake-Safe: Lessons to Be Learned from Traditional Construction ICOMOS ICOMOS*. Istanbul, Turkey. Retrieved June 13, 2011 from http://www.icomos.org/iiwc/seismic/Gulhan.pdf.

Haider, D. (2005) Incorporating disaster risk reduction into Recovery: UNDP, Bangladesh experience. *Asian Disaster Management News, 11*(2). Retrieved February 7, 2010 from http://www.adpc.net/irc06/2005/4-6/02.pdf.

Hall, N. (1997). Incorporating local level mitigation strategies into national and international disaster responses. In J. Scobie (Ed.), *Mitigating the millennium* (pp. 35–45). Warwichsire, UK: Intermediate Technology Development Group.

Hamid, S. (2015). *Bulding resilient communities through architectural interventions: Towards a sustainable future* (Doctoral dissertation). BRAC University.

Hardy, A. L., & Beeton, R. J. S. (2001). Sustainable tourism or maintainable tourism: Managing resources for more than average outcomes. *Journal of Sustainable Tourism, 9*(3), 168–192.

Havlick, S. W. (1986). Third world cities at risk: Building for calamity. *Environment, 28*(9), 6–11, 41–45.

Hillier, B., & Hanson, J. (1984). *The social logic of space*. Cambridge, MA: Cambridge University Press.

Hillier, W. R. G., Hanson, J., & Peponis, J. (1987). Syntactic analysis of settlements. *Architecture et Comportement/Architecture and Behavior, 3*(3), 217–223. Retrieved June 21, 2011 from http://discovery.ucl.ac.uk/86/1/hillier-etal-1987-synactic-analysis-settlements.pdf.

Hiwasaki, L., Luna, E., & Shaw, R. (2014). Process for integrating local and indigenous knowledge with science for hydro-meteorological disaster risk reduction and climate change adaptation in coastal and small island communities. *International Journal of Disaster Risk Reduction, 10*, 15–27.

Howell, P. (2003). *Indigenous early warning indicators of cyclones: Potential application in coastal Bangladesh* (Disaster Studies Working Paper 6). London, UK: Benfield Hazard Research Centre.

Hufstader, C. (2008). On Sandy Islands in Bangladesh, Uncertainty fills the lives of residents. *Humanitarian News*. Retrieved March 12, 2010 from http://oxfamamerica.wordpress.com/tag/floods/.

Institute for Business and Home Safety. (2009). *Hurricane Ike: Nature's force vs. structural strength*. Retrieved September 19, 2010 from http://www.ibhs.org/content/data/file/hurricane_ike.pdf.

Islam, M. S., Hasan, T., Chowdhury, M. S. I. R., Rahaman, M. H., & Tusher, T. R. (2013). Coping techniques of local people to flood and river erosion in char areas of Bangladesh. *Journal of Environmental Science and Natural Resources, 5*(2), 251–261.

Iyengar, K. (2015). *Sustainable architectural design: An overview*. Routledge.

Janif, S., Nunn, P., Geraghty, P., Aalbersberg, W., Thomas, F., & Camailakeba, M. (2016). Value of traditional oral narratives in building climate-change resilience: Insights from rural communities in Fiji. *Ecology and Society, 21*(2).

Jigyasu, R. (2002). *Reducing disaster vulnerability through local knowledge and capacity: The case of earthquake prone rural communities in India and Nepal* (Doctoral dissertation). Norwegian University of Science and Technology, Norway. Retrieved July 10, 2011 from http://lib-ezproxy.tamu.edu:2048/login?url=http://proquest.umi.com/pqdweb?did=764600331&sid=2&Fmt=2&clientId=2945&RQT=309&VName=PQD.

Johannes, R. E. (1978). Traditional marine conservation methods in Oceania and their demise. *Annual Review of Ecological Systems, 9,* 349–364. https://doi.org/10.1146/annurev.es.09.110178.002025.

Kimura, K. (1984). Ecotechniques in Japanese traditional architecture: A regional monograph of Japan. In A. Bowen & S. Yannas (Eds.), *Proceedings of the Third International PLEA Conference* (pp. 1093–1109). Mexico City, Mexico: International PLEA Organisation.

Kincheloe, J. L., & McLaren, P. (2000). Rethinking critical theory and qualitative research. In N. K. Denzin & Y. S. Lincoln (Eds.), *Handbook of qualitative research* (2nd ed., pp. 279–313). Thousand Oaks, CA: Sage.

Klein, R. J. T., Nicholls, R. J., & Thomalla, F. (2003). Resilience to natural hazards: How useful is this concept? *Environmental Hazards, 5*(1/2), 35–45.

Koçak, A., Karaca, T., Pazır, B., Zengin, B., & Kalyoncuoğlu, A. (2013). A study of the earthquake-resistance of masonry buildings constructed with local materials. *Earthquake Resistant Engineering Structures IX, 132,* 133.

Kraus, C. N., & Lin, L. (2009). Hurricane Ike along the upper Texas coast: An introduction. *Shore & Beach, 77*(2), 3–8.

Langenbach, R. (2015). *The earthquake resistant vernacular architecture in the Himalayas* (p. 83). Seismic Retrofitting: Learning from Vernacular Architecture.

Lauer, M. (2012). Oral traditions or situated practices? Understanding how indigenous communities respond to environmental disasters. *Human organization, 71*(2), 176–187.

Lebel, L. (2013). Local knowledge and adaptation to climate change in natural resource-based societies of the Asia-Pacific. *Mitigation and Adaptation Strategies for Global Change, 18*(7), 1057–1076.

Lee, A. J. (2016). *Resilience by design.* Springer.

Lefevre, M. R. (2000). Pennsylvania's historic preservation plan: Partnerships and public outreach. *Cultural Resource Management, 23*(7), 33–36.

Leis, J. (2016). Reflections on the practice of disaster resilience. In *Urban disaster resilience: New dimensions from international practice in the built environment* (p. 199).

Lewis, J. (1982). Natural disaster mitigation: Environmental approaches in Tonga and Algeria. *The Environmentalist, 2*(3), 233–246.

López-Carresi, A., Fordham, M., Wisner, B., Kelman, I., & Gaillard, J. C. (2013). *Disaster management: International lessons in risk reduction, response and recovery.* Routledge.

Mackinson, S., & Nottestad, L. (1998). Combining local and scientific knowledge. *Reviews in Fish Biology and Fisheries, 8,* 481–490.

Malhorta, S. K., & Ritchie, R. A. G. (1980). Behavior and reliability analysis of nailed timber connections under cyclic loads. In P. Karasudhi, A. S. Balasubramaniam, & W. Kanon-Nukulchai (Eds.), *Proceedings of the international conference on engineering for protection from natural disasters* (pp. 261–271). Asian Institute of Technology, Bangkok, Thailand: John Wiley & Sons.

Marshall, T. P. (2009). *Hurricane Ike damage survey.* Poster presented at the 29th Conference on Hurricanes and Tropical Meteorology, Tucson, AZ. Retrieved March 2, 2021 from https://ams.confex.com/ams/29Hurricanes/webprogram/29HURRICANES.html.

Mary, N. L. (2008). *Social work in a sustainable world.* Chicago, IL: Lyceum.

Maskrey, A. (1989). *Disaster mitigation: A community-based approach* (Development Guidelines No. 3). Oxford, UK: Oxfam International.

Maskrey, A. (1990). Supporting local initiatives in Peru. *Appropriate Technology Journal, 17*(1), 26–29.
Masterson, J. H., Peacock, W. G., Van Zandt, S. S., Grover, H., Schwarz, L. F., & Cooper, J. T., Jr. (2014). Assessing hazard exposure. In *Planning for community resilience* (pp. 71–81). Island Press/Center for Resource Economics.
McAdoo, B. G., Dengler, L., Prasetya, G., & Titov, V. (2006). How an oral history saved thousands on Indonesia's Simeulue Island during the December 2004 and March 2005 tsunamis. *Earthquake Spectra, 22*(3), 661–669.
McCorkle, C. M. (1989). Toward a knowledge of local knowledge and its importance for agricultural RD&E. *Agriculture and Human Values, 6*(3), 4–12. https://doi.org/10.1007/BF02217664.
Méheux, K., Dominey-Howes, D., & Lloyd, K. (2007). Natural hazard impacts in small island developing states: A review of current knowledge and future research needs. *Natural Hazards, 40*(2), 429–446.
Mercera, J. (2007). The potential for combining indigenous and western knowledge in reducing vulnerability to environmental hazards in small island developing states. *Environmental Hazards, 7*, 245–256. Retrieved July 10, 2011 from http://www.pacificdisaster.net/pdnadmin/data/original/Elsevier_combinedigenous_westernknowledge.pdf.
Miculax, R. M., & Schramm, D. (1998). Earthquake housing reconstruction and rural development: Joyabaj, Quiché, Guatemala. In M. B. Anderson & P. J. Woodrow (Eds.), *Rising from the ashes: Development strategies in times of disaster* (pp. 225–240). Boulder, CO and San Francisco, CA: Westview Press.
Mileti, D. S. (1999). *Disasters by design: A reassessment of natural hazards in the United States*. Washington, DC: Joseph Henry Press.
Mitchell, J. (n.d.) Protecting development. In J. Twigg (Ed.), *Development at risk?: Natural disasters and the third world* (pp. 14–19). UK National Co-ordination Committee's Working Group on the Application and Implementation of Risk Reduction Measures. London, UK. Retrieved June 29, 2011 from http://www.geomuseum.com/docs/DATRISK/DEVRISK.PDF.
Natural Hazards Research and Applications Information Center. (2001). *Holistic disaster recovery: Ideas for building local sustainability after a natural disaster*. Boulder, CO: University of Colorado [Updated February 8, 2005]. Retrieved on April 15, 2003 from http://www.colorado.edu/hazards/holistic_recovery/.
Office of the United Nations Disaster Relief Coordinator. (1982). *Shelter after disaster: Guidelines for assistance*. New York, NY: United Nations Disaster Relief Organization. Retrieved July 2, 2011 from http://reliefweb.int/sites/reliefweb.int/files/resources/E4FE896AFFF16709C1256CB10056558E-undro-shelter1-jul82.pdf.
Pearce, T., Ford, J., Willox, A. C., & Smit, B. (2015). Inuit traditional ecological knowledge (TEK), subsistence hunting and adaptation to climate change in the Canadian Arctic. *Arctic, 68*(2), 233.
Pfeifer, K. (1996). Modernization and indigenous knowledge: Ideas for development or strategies of imperialism? *Peace and Change, 21*(1), 41–67. https://doi.org/10.1111/j.1468-0130.1996.tb00254.x.
Phillips, B. D. (2015). *Disaster recovery*. CRC press.
Poncelet, J. L. (1997). Disaster management in the Caribbean. *Disasters, 21*(3), 267–279. https://doi.org/10.1111/1467-7717.00061.
Rapoport, A. (1969). *House form and culture*. Englewood Cliffs, NJ: Prentice-Hall.
Rawal, V., & Desai, R. (1994). *Repairs and retrofitting: A manual for people of Marathwada*. Ahmedabad, India: Ahmedabad Study Action Group (ASAG).
Rogers, P. (2016). *Resilience & the city: Change,(dis) order and disaster*. Routledge.
Rothrock, G. C. (2000). A tale of two communities: Community involvement in preservation planning. *Cultural Resource Management, 23*(7), 23–26.
Rungamanee, S., & Cruz, I. (2005). The knowledge that saved the sea gypsies. *A World of Science, 3*(2), 20–23. Retrieved January 11, 2011 from http://portal.unesco.org/science/en/files/4527/11158120231Sea_gypsies.pdf/Sea_gypsies.pdf.
Sapre, P. M. (2000). Realizing the potential of management and leadership: Toward a synthesis of western and indigenous perspectives in the modernization of non-western societies. *International Journal of Leadership in Education, 3*(3), 293–305.

Schilderman, T. (1993). Disasters and development: A case study from Peru. *Journal of International Development, 5*(4), 415–423.
Schilderman, T. (2004). Adapting traditional shelter for disaster mitigation and reconstruction: experiences with community-based approaches. *Building Research and Information, 32*(5), 414–426.
Schilderman, T. (2016). How does reconstruction after disaster affect long-term resilience? In *Urban disaster resilience: New dimensions from international practice in the built environment* (p. 120).
Schmuck-Widmann, H. (2001). *Facing the Jamuna River: Indigenous and engineering knowledge in Bangladesh.* Bangladesh: Bangladesh Resource Centre for Indigenous Knowledge.
Shah, H. (2003). *The last mile: Earthquake risk mitigation assistance in developing countries.* Stanford, CA: Stanford University.
Singhal, R. (2000). A model for integrating indigenous and scientific forest management: Potentials and limitations for adaptive learning. In A. Lawrence (Ed.), *Proceedings of the forestry, forest users and research: New ways of learning workshop* (pp. 133–140). Vienna, Austria: European Tropical Forest Research Network.
Smith, G., & Wenger, D. (2006). Sustainable disaster recovery: Operationalizing an existing framework. In H. Rodriguez, E. Quarantelli, & R. Dynes (Eds.), *Handbook of disaster research* (pp. 234–257). New York: Springer.
South Pacific Applied Geo-Science Commission. (2004). *Implementing the Yokohama strategy and plan of action: Pacific Islands regional progress report (1994–2004).* Suva, Fiji: South Pacific Applied Geo-Science Commission.
Spennemann, D. H. R. (1998). *Non-traditional settlement patterns and typhoon hazard on contemporary Majuro Atoll, Republic of the Marshall Islands.* Albury, Australia: Charles Sturt University. Retrieved July 2, 2011 from http://marshall.csu.edu.au/Marshalls/html/typhoon/typhoon.html.
Subasinghe, C. (2013). Spatial confrontations: Abandonment of self-labor in transitional sheltering after a natural disaster. *International Journal of Disaster Risk Reduction, 6,* 78–86.
Takeyama, K., Hisada, T., & Ohsaki, Y. (1960). Behavior and design of wooden buildings subjected to earthquake. *Proceedings of the Second World Conference on Earthquake Engineering, 3,* 2093–2111.
Tirpak, S. T. (2009). United States Army Corps of Engineers: Galveston district operational experiences and response to Hurricane Ike. *Shore & Beach, 77*(2), 60–70.
Twigg, J. (2004). Disaster risk reduction: Mitigation and preparedness in development and emergency programming. *Good Practice Review, 9.* Retrieved June 29, 2011 from http://www.odihpn.org/report.asp?ID=2618.
Twigg, J., & Bhatt, M. (Eds.). (1988). *Understanding vulnerability: South Asian perspectives.* London, UK: ITDG Pubishing.
Twigg, J. (n.d.) Disasters, development, and vulnerability. In J. Twigg (Ed.), *Development at risk: Natural disasters and the third world* (pp. 2–7). UK National Coordination Committee's Working Group on the Application and Implementation of Risk Reduction Measures. London, UK. Retrieved June 29, 2011 from http://www.geomuseum.com/docs/DATRISK/DEVRISK.PDF.
United States Geological Survey. (2008). *Coastal change hazards: Hurricanes and extreme storms.* Retrieved May 21, 2010 from http://coastal.er.usgs.gov/hurricanes/gustav/coastal-change/.
Veitayaki, J. (2002). Taking advantage of indigenous knowledge: The Fiji case. *International Social Science Journal, 54*(3), 395–402.
Walshe, R. A., & Nunn, P. D. (2012). Integration of indigenous knowledge and disaster risk reduction: A case study from Baie Martelli, Pentecost Island, Vanuatu. *International Journal of Disaster Risk Science, 3*(4), 185–194.
Weichselgartner, J., & Obersteiner, M. (2002). Knowing sufficient and applying more: Challenges in hazard management. *Environmental Hazards, 4*(2–3), 73–77.
Wisner, B., Blaikie, P., Cannon, T., & Davis, I. (2004). *At risk: Natural hazards, people's vulnerability and disasters* (2nd ed.). London, UK: Routledge.
Yates, L., & Anderson-Berry, L. (2004). The societal and environmental impacts of cyclone Zoë and the effectiveness of the tropical cyclone warning systems in Tikopia and Anuta, Solomon Islands, 26–29 December 2002. *Australian Journal of Emergency Management, 19*(1), 16–20.

Chapter 4
Tensegrity Model

> The way community builds tells you, sometimes, all you need to know about its values: just to look at Radburn, New Jersey, will tell you that it is a suburb built to control the automobile, in the same way that it does not take long to figure out that Positano and the rest of the Amalfi Coast in Italy were built to connect to the sea. (Goldberger, 2009, p. x)

This chapter processes ways of customizing theories and approaches conducive to the discourse's primary objective of grounding long-term disaster resiliency through sustainability that is embodied in tacit knowledge. This is to be understood as a quasi-empirical demonstration of a possible relationship between spatial integration and disaster resiliency as portrayed in hurricane-prone spontaneous settlements in Bolivar Peninsula, Galveston County, Texas. The tensegrity analytical model deployed in this demonstrative methodology piece appropriates structured observations based on photographic evidence, statistical analyses on damage assessment and syntax studies of flood permeability to reveal potential for disaster researchers to elevate the discourse beyond the rhetoric. While structured observations, statistical analyses and syntactical studies make the threefold structural backbone of the methodological model, the equilibrium or rigour entirely dependent on the tensional integrity achieved through the specific tools: binary regression, variance analysis, cross tabulation, ariel views, elevation views, structural component studies, mapping studies and axial connectivity calibrations.

The study explores environmental and economic sustainability in terms of innate disaster survival in coastal settlements consisting of self-built or locally built houses that were built by small-scale local contractors, sometimes identified as carpenters, artisans, or homebuilders. The selection of settlements with specific types of dwellings was based on the understanding that local technology and skills are best suited the local environmental conditions, including extreme natural events specific to the area such hurricanes. The chapter also examines the survival strategies of coastal settlements for surge or flood vulnerability in terms of permeability of domestic units and spontaneity of settlement organization. This is based on the premise that speedy recovery during the post-disaster phase is based upon many locally manageable factors including readily available local expertise, infrastructure, and local materials that are amply available at an affordable cost for both owners and donors.

C. Subasinghe, *Rebuilding for Resilience: A Barrier Island Case*,
https://doi.org/10.1007/978-3-030-65532-7_4

First, syntactical methods were used to identify spatial patterns connected with surge exposure. One such pattern, the visibility or axial maps, demonstrates integration, permeability, and connectivity among settlements. Second, based on these spatial maps, the characteristics of the surviving residential built fabric was described using a rubric of architectural and structural make-up. The key research question leading to the proposed research design that has been handled within the broader context of different causes and types of hurricane damage is detailed in the final portion of the chapter.

Ecological Inquiry

The purpose of this model study was to demonstrate an integrated, dynamic perspective of how local knowledge and disaster resilience influence each other. In particular, the study investigated settlements located on Bolivar Peninsula, which featured local knowledge and highlighted the capacity of coastal communities for mitigation, preparedness, and recovery from hurricanes. First, there are specific qualities embodied in how settlements are laid out and how buildings are set in relation to an access grid, and how local technology, skills, materials, and labor are utilized. Second, these physical attributes provide an in-depth understanding of the spatial features that make residential units and settlement less susceptible to surge damage (typology of built form, accumulation of various built layers of different ages, topographical determinants). In addition, the study analyzed the impacts of the stated spatial features on local "know-how" used in managing pre-disaster vulnerability to hurricanes. Third, the settlements located on the southwest part of Bolivar Peninsula were closest to the path that Hurricane Ike took as it crossed the Gulf of Mexico and Galveston Bay, while the settlements on the northeast part of the island were farthest away. Furthermore, exploiting the rectilinear nature of the coastal settlements on Bolivar, with its direct relationship to Hurricane Ike's path, is critical to understanding hurricane damage and survival probability relationships. The theoretic pragmatics behind the study that oversees sustainable post-disaster recovery is twofold, with hurricane damage being minimal due to in-built resiliency of the spontaneous settlements and the damage being easily and locally addressable due to readily available local capacity and infrastructure.

Premises, Questions, and Objectives

The premise of this demonstrative piece is based on the understanding that spontaneous settlements including disaster-prone communities are sustainable when they can survive and progress through major recurring-natural events. These traditional communities are sustainable as a way of life. According to the World Food and Agriculture Organization (2009), generational "know-how" and networks of communities and households can be mobilized to prepare for, mitigate, and manage disasters related to climate change. By dismantling the grand concept of physical context for disaster-vulnerable spontaneous settlements, one can potentially ascertain sustainable disaster resiliency and recovery built into the planning, construction, and operational mechanisms of the settlements. Changes made to structures over time for various operational reasons such as maintenance, progressive development, and other various functional mechanisms may be important side effects of disaster resiliency for such settlements. This study tried to determine if a probable connection existed between spatial integration and disaster resiliency among hurricane-prone spontaneous settlements on Bolivar Peninsula and Galveston Island in Texas. The study also explored the environmental and economic sustainability of these settlements in terms of exploiting the hard-wired disaster survival ability of spontaneous settlements as an instigator of rehabilitation and reconstruction efforts. The selection of settlements with specific types of dwellings (either self-built or locally-built) was based on an understanding that local technology and skills are best suited to address the socio-economic and environmental conditions of an area, even in extreme natural events such as tornados, hurricanes, and tropical storms. The study also zooms into survival strategies used by these settlements to deal with recurring hurricanes, with special reference to the nature of structural elevation above existing ground level, and spatial organization of the overall settlement in relation to the access grid. As stated previously, uncovering these survival strategies was based on the premise that speedy recovery during the post-disaster phase in part relies on easily adaptable resiliency conducive to local expertise and infrastructure.

The operational question asked how spatial configuration, with regard to permeability, affected the performance of selected neighborhoods on Bolivar Peninsula during Hurricane Ike. This is based on the hypothesis that the configuration of the grid has a primary function in determining the permeability of a settlement, because higher permeability provides more unobstructed escape routes and options, and surge action associated with property damage has a negative correlation to disaster vulnerability. The hypothesis further predicted that the elevated structures in the indicated settlements were more permeable than non-elevated ones as well as being less vulnerable to flood damage. A secondary question generated by the study investigated hurricane-resistant built characteristics of single-family detached houses. The key to this question was the observation that certain types of architectural characteristics, such as those found in the "Last House Standing," are less susceptible to hurricanes than the other characteristics (Fig. 4.1). Therefore, an analysis of the architectural typology of surviving structures helped answer the vital part of the operational question with regard to local expertise versus disaster resiliency.

The following complimentary questions to the hypothesis are intended to ascertain an in-depth understanding of concrete and discrete attributes attached to surge damage via a settlement matrix and permeability rubric.

1. What is the relationship of each level of hurricane damage (no damage, minor, moderate, major, and destroyed) to the axial connectivity of the access grid?
2. How do the two extreme ends of the property survival matrix (no damage and destroyed) vary along the connectivity gauge?

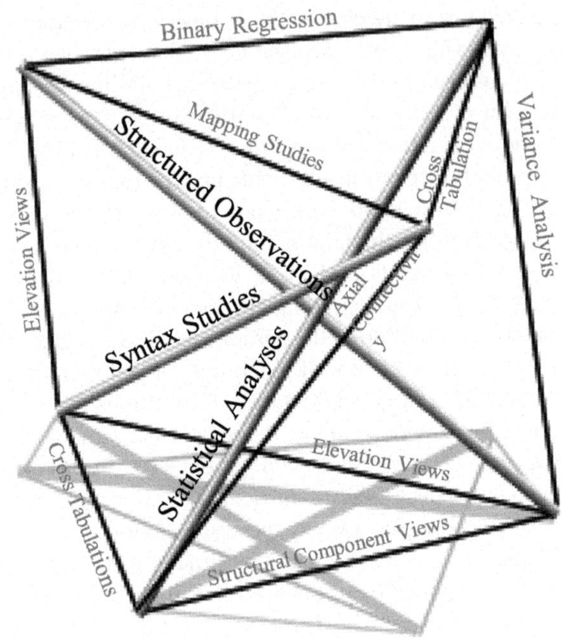

Fig. 4.1 Tensegrity methodological model

3. Which of the characteristics of structural elevation found among the residential units deals directly with surge permeability?
4. Which typologies or combinations of built characteristics best represent the architectural make-up of the surviving structures?

Representative Cases

The distinctive need for a case study surfaced out of the necessity to investigate the apparent, yet complex phenomena linked to the potential relationship between one of the key physical attributes of disaster-prone spontaneous settlements and disaster vulnerability. Spatial integration of settlements that were not built as "designed to sell" in current real estate practice was the main consideration in selecting an appropriate case for this study. The Gulf of Mexico shoreline defines at least one of the boundaries for all the settlements analyzed in this investigation. A few settlements located on High Island have no single coastal boundary and the only landward settlement that is located on the mainland adjoins Chambers County.

Bolivar Peninsula was named after Simón Bolívar (1783–1830), the South American hero. It is a narrow stretch of land known as a barrier island. It extends 27 miles along the Texas Gulf Coast and stretches to the northeast to form part of eastern Galveston County. The center of the peninsula is located at 29°26’ N, 94°41’ W (Daniels, 2010, para. 1). The widest point of Bolivar Peninsula is between Crystal Beach and Caplen and measures three miles. Its narrowest point, where Rollover Pass segments the Gilchrist area into two districts, is just one-quarter mile. The causeway between Bolivar Peninsula and Galveston Island is less than three miles long. Ground access from the Texas mainland to Bolivar Peninsula is restricted to southern Chambers County. The sheltered Gulf Intercoastal Waterway along the north side of Bolivar Peninsula provides freight transportation as well as creating a marine entrance from the Gulf of Mexico to Galveston Bay. Although Crystal Beach is the only incorporated area, Bolivar Peninsula also includes the unincorporated communities of Caplen, Gilchrist, Port Bolivar, and High Island. Not only does Bolivar Peninsula provide a tangible sample of spontaneous settlements that face recurring hurricanes, it also provides a manageable scale for settlements' age, damage severity, disaster statistics accessibility, and time lapse between pre and post-disaster reconstruction.

The following rationale was used in the final selection of Bolivar Peninsula as a suitable case to investigate the dichotomy between damage and resiliency.

- Bolivar Peninsula, located directly along a common path for hurricanes forming in the Gulf of Mexico, was severely affected by Hurricane Ike in 2008.
- Most of the housing on Bolivar Peninsula was built before 1980 in neighborhood subdivisions, but prior to the introduction of developer-built tract housing. According to the 2000 census and statistics obtained from US Census Bureau (2003), 30% of the housing block was constructed before 1959.
- Older residential areas on Bolivar Peninsula demonstrate typical vernacular characteristics of East Texas architecture and more spontaneously acquired settlement patterns original to early Spanish cities, towns, and hamlets.
- Information on levels of damage to randomized properties, physical characteristics, and household data was obtained by Texas A&M University's Hazard Reduction and Recovery Center after Hurricane Ike in 2008 and updated in 2010.
- Verification data for post-disaster property status was available from appraisal cards maintained by the Galveston County Appraisal District.

The Tensegrity Components Explained

The descriptive analysis of precedent and case studies through existing literature revealed a distinct need for investigating the relationship between local knowledge and surge damage. It supported the existence of such a relationship for seismic and flood prone vernacular settlements in many parts of the world. Therefore, the objective scrutiny of both the micro aspect of spatial fabric, including individual spaces as well as a network of open spaces, and the micro aspect of the spatial fabric that constitutes the individual settlements and their morphological interactivity are pertinent to this study. Thus, the study handles its objective evaluation of *place* at the settlement or neighborhood scale and *space* at the individual house scale in Bolivar Peninsula. Field observations were used for both place and space investigations and the techniques used were space syntax and permeability analysis. These techniques were executed through damage assessment form and elevation characteristic table/rubric respectively.

The interpretation of spatial elements requires an abstract and objective frame of thought, quantifiable data, and ideally, the language of mathematics (Tuan, 2001). As basic elements of location-based investigations, place and space are used to derive necessary constructs for spatial description. However, place has more substance than geographical demarcation; it is a unique entity or special ensemble that has history and meaning (Lukermann, 1964; May, 1970; Tuan, 2001). On the other hand, space provides cues for our behavior, which varies with the individual and cultural group (Downs, 1970; Hall, 1966). According to Tuan (2001), geometrical space is cultural space, a sophisticated human construct, the adoption of which has enabled us to control nature to a degree hitherto impossible. Figure 4.2 condenses the research design into three distinct territories. First are the broad domains of vernacular, indigenous, spontaneous, folk, or local styles with their continuously evolving boundaries. These may consist of rural, urban, residential, and commercial precincts. Second is the objective evaluation of place (settlements) through the syntactical analysis. Third is the objective evaluation of the space (houses) within place (settlements) via the permeability rubric.

Appropriated for this inquiry with a broad definition of vernacular, the demonstrated methodology uses multiple methods of data generation to formulate a hybridized database with rich scientific and empirical evidence. For example, the structured observations carefully document a combination of machine and human-generated data sets: ArcGIS maps and hand-drawn settlement lay outs, photographic studies of elevation characteristics and freehand drawing studies of houses based on field sketches and photographs taken before and after the studied hurricane.

Fig. 4.2 The structure called "Last House Standing" in Gilchrist was built by the owners after Hurricane Rita (*Source* IBHS, 2009)

Objective Evaluation of Space

In order to evaluate space, this study examined the elevation characteristics of hurricane-prone Port Bolivar houses. The basis of this inquiry was the understanding that vernacular architecture cannot be interpreted holistically apart from its contents and its context. Interpreting it requires histories of vernacular architecture that integrate furniture, yards, farmsteads, and ultimately, settlement patterns into the whole. On the other hand, it is not difficult to find arguments that describe materials related to vernacular architecture as more valuable than other kinds of architecture for understanding "most people" because of its embedded communal values. However, for vernacular architecture and other material culture scholarship, this rather intriguing quality of objects has too often resulted in essays that have been purely descriptive (Upton, 1983). Due to less elaborate structural sophistication, architecture of the common folk has not survived as have houses owned by the wealthier segments of the population. In most cases, the types of vernacular architectural pieces are those that not only passed the test of the time, but also maintained their numbers and have accommodated the lifestyles of successive generations. The definition of vernacular architecture has grown to cover many different types of architecture. It now includes anything that has not been included in mainstream architectural history, including nineteenth or twentieth century speculative development, industrial buildings, fast-food restaurants, and other commercial franchises. In other words, virtually anything not obviously the product of an upper class, avant-garde, aesthetic movement is often seeing as vernacular. The idea of commercial vernacular architecture was added in early twentieth century (Venturi, 1977).

According to Upton (1983), many scholars prefer to define vernacular architecture not as a category that includes certain types of buildings, but as an approach to architectural studies, which enables traditional architectural historical inquiries. Upton also identified four basic approaches to vernacular architectural studies: (a) object oriented studies, (b) socially oriented studies, (c) culturally oriented studies, and (d) symbolically oriented studies. This study takes the object-oriented or object-specific approach to examine human alertness to recurring environmental phenomenon and responsiveness to extreme climatic conditions. The process analyzes both settlements and dwellings for clues of such responsiveness. Even though the current practice of modern object-oriented studies has gone far beyond antiquarianism, scientific or otherwise, it still relies somewhat on intuitive, rather than explicit, concepts of change. Because artifacts are products of human creativity, even very basic antiquarianism makes certain rhetorical guesses about the authors and users of those artifacts.

As Upton (1983) posited, instead of relying on such guesses, object-oriented researchers typically rely on a kind of common-sense functionalism and on aesthetic trickle-down theories to account for architectural choices. In contrast, many researchers who find that exploitation of architecture is more reliable in understanding its authors and users dismiss the rhetorical predicting of the people who live with a view to understanding their artifacts. This reversal of examining objects as evidence about past and present human transactions with their socio-physical environment has been identified as the single most important turning point in the study of vernacular architecture in the last few decades.

Objective Evaluation of Place

In order to evaluate place, this demonstrative case obtained data on connectivity among open space network/access grid by utilizing Space Syntax. These series of network-grid spaces are represented by a map of interconnected or segmented axial lines. The connectivity is acquired as a measure of relative positioning (connectivity) of an axial line to other axial lines in the network or grid. Space syntax was proposed by Hillier and Hanson (1984) as a starting point for examining spatial configurations in built forms and fabric. Gamma analysis or access analysis of space syntax provides a system to identify not only how spaces within a structure are arranged and related to each other, but also how a particular built form mediates the relationship between its occupants and visitors. Furthermore, while highly integrated areas tend to attract movement and often function as public domain, peripheral or poorly connected areas are used primarily as private spaces due to their poor accessibility (Bafna, 2003; Hillier, 1996, 2004; Hillier, Penn, Hanson, Grajewski, & Xu, 1993). Obtaining accessibility is possible through translating a building into an access graph or gamma map, in which each room is represented as a circle; the access between rooms is represented as lines linking circles together. A justified access graph is used to represent rooms that are the same depth from the outside on the same level.

The basics of the space syntax are such that it presupposes structure of the space around buildings; according to the syntactical theory such spaces are structured in a way that allows strangers to move about, but only inhabitants and certain visitors are allowed inside structures. Because the inhabitants are the controllers, they have invested in power over their settlements and dwellings; in contrast, visitors enter or stay as subjects of the system as the controlled (Markus, 1993). Therefore, while most interior spaces of a building are occupied by the inhabitants, visitors occupy the outer or peripheral casing of the building. However, syntactical patterns should be analyzed though the filter of the particular people using that space, an element that is both culturally specific and difficult to determine without living informants (Morrow, 2009).

The role of space syntax in this study was to provide a basis to identify nodes of settlements that have survived hurricanes, and further analysis is required to describe elements of the built environment that demonstrate resiliency. Rapoport (1990) averred that such characteristics are demonstrated by the use of fixed-feature elements like wooden walls, columns, thresholds, and plastered floors, and semi fixed-feature elements such as doors, benches, hearths, wells, and other furnishings and portable artifacts. In order to capture the spontaneity of urban space, it is essential not to lose the inherent peculiarities of the system. Hillier and Hanson's space syntax approach has been utilized to acquire both consistent and objective data for spatial relations unique to settlements (Hillier, 1996; Hillier & Hanson, 1984). Axial and convex maps were drawn of sleeted there settlements to extract the spatial pattern.

The maximum global linear extension of any segment of the system inside the settlement can be distilled from the axial map. On the other hand, convexity, the counterpart of axiality, represents the two-dimensional extension of space. The global and local aspects of space syntax are based on the argument that every point in the system inherits both a one-dimensional and a two-dimensional form. Socially interpreted, while convex space illustrates where the person is in the open space system, axial lines offers the details of the location where she/he might be going. This is related to the notion that axiality is a measure of movement and convexity is more about mutual presence. Therefore, axiality is associated predominately with strangers, while the common denominators of convexity are related to inhabitants (Darjosanjoto, 2007; Hillier, Hanson, & Peponis, 1987).

For the purpose of this analysis, a grid of an open space network was drawn with CAD and was based on GIS maps and aerial photos from Google Earth for Bolivar settlements. DepthMap generated axial maps were created using these open space maps as base drawings. Figure 4.3 shows Bolivar settlements at either side of the Gulf of Mexico Intercoastal highway and land parcels randomized for damage assessments. The methodology used an axial line analysis to derive syntactical measures of connectivity using DepthMap as detailed below:

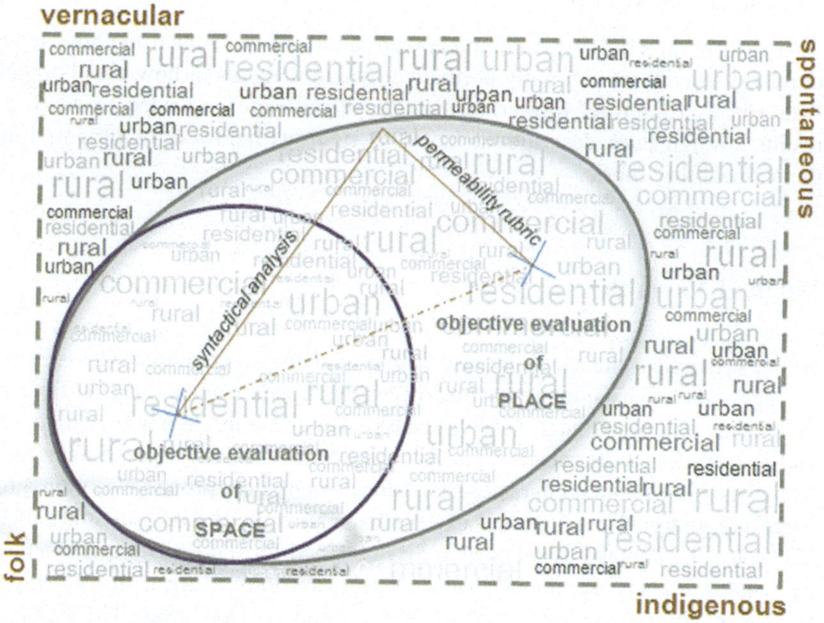

Fig. 4.3 Research design

1. Connectivity is a measure of relative standing of an axial line in terms of its position in a series of interconnected axial lines. In the broadest sense, connectivity is the number of streets or open spaces represented by axial lines directly connected to a particular street or open space represented by an axial line. The current study was concerned primarily with axial data; however, one of the selected settlements for ground-elevation analysis was also analyzed using convex maps.
2. Connectivity is a measure of relative standing of an axial line in terms of its position in a series of interconnected axial lines. In the broadest sense, connectivity is the number of streets or open spaces represented by axial lines directly connected to a particular street or open space represented by an axial line. The current study was concerned primarily with axial data; however, one of the selected settlements for ground-elevation analysis was also analyzed using convex maps.
3. Integration is a measure of accessibility among routes presented by axial lines with two levels of measures, global and local integration. Global integration also refers to integration radius max; this represents numerically the average number of turns needed to get to a particular route from any other route in the network. Furthermore, global integration measures the accessibility of a route represented by an axial line from the city as a whole. From a pedestrian circulation standpoint, global integration predicts how easy it would be for a stranger to find a specific street. In contrast, local integration measures the connectivity of a street represented by an axial line to another axial line or lines that are *n* steps farther from the referred line.

It is to be noted that the computational nature of syntactical methods tends to draw too heavily on culturally normative assumptions about how people organize their space. Therefore, further verification of outcome is needed to confirm the results through thick descriptions based on rigorous fieldwork as proposed under the permeability rubric.

Verified Variables

In the syntactical analysis, because levels of damage are observed assessments, average connectivity among axial lines becomes the independent variable. The dependent variable, surge damage, has five levels of damage: none, minor, moderate, major, and destroyed. In the permeability analysis, the categorical/nominal variables fall into five groups of elevation characteristics: street orientation, ground elevation, ground enclosure, external bracing, and external attachments. The current study distilled Hurricane Ike damage assessment values obtained for the Bolivar settlements into axial maps. The average damage values for grid sections or streets in each neighborhood were assigned to fewest line maps generated via DepthMap. The significance was evaluated with surge damage as one of the three contributors to structural or overall damage: wave action, surge, and storm. The statistical procedure handles two types of data: (1) five levels of damage (none, minor, moderate, major, and

Table 4.1 Matrix of damage assessment and permeability characteristics

Survival	Damage level	Elevation characteristics
Yes	None	Street orientation
	Minor	Ground elevation
No	Moderate	Ground enclosure
	Major	External bracing
	Destroyed	External attachment

destroyed) against the axial connectivity of Bolivar Peninsula, and (2) survival binaries (survived, destroyed) against five categories of elevation characteristics (see Table 4.1).

Syntactical Analysis

The average connectivity for each segment of the grid against the level of damage to residential units (houses) may reveal possible relationships between dependent and individual variables. A comparison of means of variations to find significance among outcome and predictor variable can be done using a particular inferential statistical method, one-way Analysis of Variances (ANOVA), among five levels of damage. In addition, building up a typology for hurricane-resilient architectural characteristics for all the settlements is needed to fill the "local know-how" portion of the operational question. This was instigated by forming a permeability rubric based on the built features pertaining to structural elevation above the exiting ground level of residential properties that have been identified as positive towards surge permeability in the post Hurricane Ike literature.

A binary logistic regression is appropriate for determining which characteristics or combinations of characteristics of the permeability rubric work best against surge damage. This analysis limits discussion of the details of the mathematics involved and the statistical basis of the space syntax analysis to the basic principles above. This is an acknowledgement of its importance, as well as an acknowledgement of the author's lack of familiarity with the inner workings of this method using a computer program. The DepthMap software, which was developed by the University College London, requires simplified plan drawings of structures in order to process various forms of spatial analyses to depict simple graphics that group value ranges along a color-based scale. This allows for intuitive visual comparisons of spatial variables followed by regression analyses, which will be explored in detail in the next chapter. Fundamentally, the methods and rational of space syntax analysis presents itself to architects and planners as an incredible tool with great potential to clarify the internal dynamics of spontaneous settlements (Morrow, 2009).

Permeability Rubric

The physical makeup of residential units that alter the fluid dynamics of surge flow critically determines permeability. Houses elevated on columns and free of first/ground floor walls may provide the best scenario for allowing surge to escape from high to low pressure zones as well as limiting water damage to cosmetic repairs, or at worst, replacement of certain structural members. However, no matter how permeable the structures are, unsound construction will not survive all three types of hurricane damage (wind, wave, and surge). The rubric is based on five broad categories of external built characteristic that were identified in post Hurricane Ike literature as key players in surge permeability (see Table 4.2). The following criteria were adopted for use in distilling many elements under each category into four representative models. The rationale used for each permeability group in deciding the scope of different characteristics is outlined in the following sections. The permeability rubric developed on various elevation characteristics emphasized in Hurricane Ike literature focuses on analyzing structural makeup pertinent to surge damage assessment of Port Bolivar settlement..

Orientation

The properties located in the intersection or Conner plots were categorized into a model labeled "access road width is greater than offset from access road." Access road offset and width was considered for the properties with similar length and width dimensions (square in shape): *if* the unit is a corner property, *then* Offset from access road is less than road width. If the width of the access road is approximately close to offset from the access road, either "oriented length wise to access road" or "width wise to access road" was adapted: *if* the access road equals offset from the access road, *then* either lengthwise or width wise.

Elevation

As interiors were not accessible for detailed inspections, the houses with solid concrete, stone, or brick plinth were categorized under "boarded plinth" category. The average story height was based on residential units' floor to ceiling height as stated in International Code Council (ICC) (2011).

Enclosure

Because freestanding columns are also braced at the second floor level, exposed wooden frames have been considered to behave as similar to them too. The temporary partitions on first/ground floor made during the hurricane season are not being included in the Enclosure group.

Table 4.2 Elevation characteristics of the permeability rubric

Street orientation	Ground elevation	Ground enclosure	External bracings	External attachments
1. Lengthwise from access road 2. Widthwise from access road 3. Offset from access road < road width 4. Offset from access road > road width				
	1. Boarded plinth 2. Exposed plinth 3. Column height > average story height 4. Column height < average story height			
		1. Open with columns 2. Columns with a room 3. Columns with rooms 4. Enclosed with rooms		
			1. No bracings 2. Braced at one direction 3. Braced at opposite directions 4. Braced at multiple directions	
				1. No attachments 2. Front/rear patio/deck 3. Side patio/deck 4. Wrapping veranda

Table 4.3 Specimen photographic analysis of permeability rubric

Elevation characteristics - permeability rubric					
	Ground Orientation	Ground Elevation	Ground Enclosure	External Bracing	External Attachment
1					
2					
3					
4					

Bracing

Only the devices those appeared to be contributing to structural integrity of first/ground floor vertical supports have been considered. Even for the properties with no columns but all walls, bracing was considered as some of them had exposed wall bracings crisscrossing column struts.

Attachments

Temporary sheds, semi-detached garages and workshops, and mobile storages attached to main house have not been considered in this group. In addition to residential units with continuing verandas and three or more facades, the units that constituted interconnected patios or decks on more than two sides counted as wrapping verandas.

The detailed descriptions of the elevation characteristics shown in Table 4.3 were obtained through on-site observations. These photographic evidence was further verified using Galveston county appraisal cards.

Selective Sampling

Because this study analyzed the socio-cultural production of human habitation, it employed a higher level of awareness due to its subjective normatives. Settlements were selected based on three parameters: the age of the settlement with special reference to a number of self-built or locally built units in each settlement, distance from the hurricane path, and access to damage assessment data. The oldest neighborhood was closest to the hurricane path on the edge of the Bolivar side of Galveston Bay (Port Bolivar). In contrast, the comparatively new neighborhoods were farthest from the hurricane path facing the Caribbean Sea side of the Gulf of Mexico (see Fig. 4.4). Except for High Island, most of the landward settlements were situated approximately in the middle of the coastal settlements between Port Bolivar and Bolivar Peninsula. Damage data for Bolivar Peninsula was collected as part of the Galveston Appraisal District Hurricane Ike damage assessment survey, produced by the Hazard Reduction and Recovery Center (HRRC) of Texas A&M University's College of Architecture. Random sampling and ArcGIS were used to generate property IDs for single-family detached houses. Specific residential properties were excluded from the sample, based on the selection criteria established in the hypothesis: (a) new construction and completely remodeled houses built after Hurricane Rita in 2005, (b) recreational vehicles (RVs), mobile homes, and replacement housing, and (c) ocean properties (houses built in the ocean). The population considered for the one-way Analysis of Variance (ANOVA) was Bolivar Peninsula. For the logistic regression, an unknown settlement (Port Bolivar) on the island was cluster sampled (see Fig. 4.5). In each cluster, all the houses were sampled (one stage cluster sample) (Fig. 4.6).

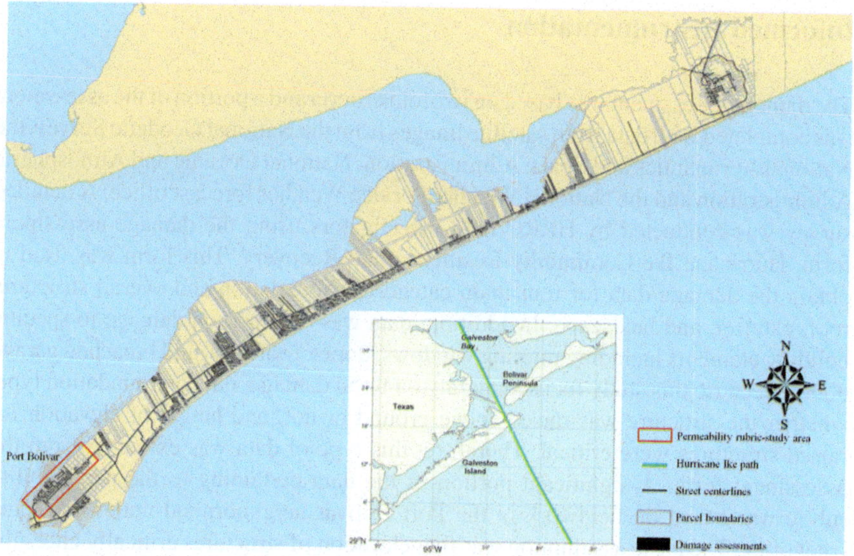

Fig. 4.4 Hurricane Ike path and ArcGIS randomized land parcels for damage assessments (*Source* Doran, Plant, Stockdon, Sallenger, & Serafin, 2009)

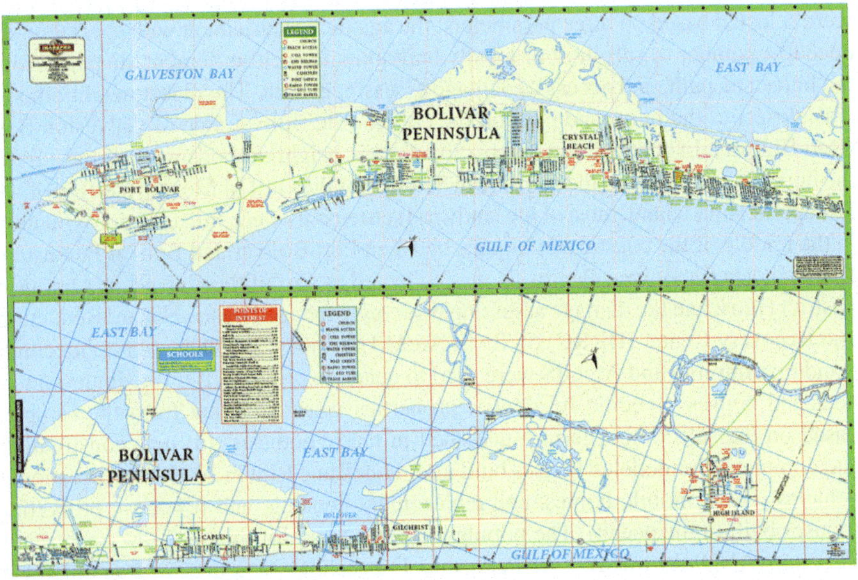

Fig. 4.5 Land use map showing left and right halves of Bolivar Peninsula (*Source* Bolivar Chamber of Commerce, 2010)

Informed Instrumentation

The damage assessment was based on twofold criteria and a portion of the assessment was completed with data from satellite images from the National Geodetic Survey, the National Aeronautics and Space Administration, National Oceanic and Atmospheric Administration and the National Weather Service Weather forecast office. A detailed survey was conducted by HRRC field investigators using the damage assessment form, Hurricane Ike Community Resilience and Recovery. This form was used to obtain the damage data for four main categories: foundation and overall structure, roof, exterior, and landscape. In addition, data was obtained on damage to specific building elements including missing windows, doors, roof tiles, and breached garage doors. Because this study focused mainly on flood damage, data on foundation type, whether the structure was raised above ground or not, and height of elevation for raised structures were critical. Therefore, this type of data was extracted from the assessment form. A significant portion of the data pertaining to the permeability rubric was based on fieldwork in the Port Bolivar neighborhood under five main categories. The basic assumption was that elevation of structures critically alters the surge damage by means of permeability. Galveston Central Appraisal District cards were used to cross check the dimensions of building footprints, foundation details, age of the structures, attached and non-attached secondary structures, and plot size against official observations on permeability characteristics.

Fig. 4.6 ArcGIS generated property IDs of the oldest "permeability rubric settlement" in Port Bolivar

Data Collection

The data collection in January 2010 for each randomly selected household was done only when the field investigators were granted permission to photograph and assess the structure. Therefore, assessment and photographing were restricted to streets in the absence of occupants' consent. The field investigators were asked to photograph the house number, if present, in order to confirm the Galveston County Appraisal District property IDs and location. However, satellite images/aerial photographs were utilized to assess the inaccessible properties. Piling up of debris barricading access, failures to obtain consent for access, and difficulty in identifying the location due to partial or complete destruction of the property were a few of the reasons for inaccessibility. The permeability rubric based final set of data on Port Bolivar neighborhoods relied on Galveston District appraisal cards for verification on building footprints and accompanying structures. However, additional information including external bracings and ground enclosure details was gathered on site using field observations in 2010. Wall heights were based on standard siding sizes and column heights accordingly.

Data Analysis

Representation and analysis of an open space structure or system of open spaces of a settlement presents numerous morphological constrains, including spatial continuity. Even though the majority of interiors of buildings are subdivided into a series of discrete units or compartments (rooms), the exterior or outside has no such distinguishable compartmentalization. Hillier and Hanson's (1984) and Hillier et al.'s (1987) work on syntactical analysis allows one to assess systems of exterior open spaces via axial and convex maps. In order to evaluate elements of environmental responsiveness embedded in spatial artifacts objectively, an understanding of how each space is integrated with the rest of the building is critical.

According to Hillier and Hanson's (1984) and Hillier et al.'s (1987), this can be achieved using the access graph of a building to calculate the following steps:

a. Control Value (CV) measures the degree of connectivity of a particular space to its immediate neighbors. For calculation purposes, each space in the building is assigned a value of 1; this value is divided among each of the connected-adjacent spaces. Added together, the resultant value presents the control value for each space. The degree of control increases with higher CV numbers.
b. Mean Depth (MD) measures the depth of a particular space in relation to the other spaces in the building. MD = cumulative depth of each space/$p - 1$, where p is the number of points in the system.
c. Total Depth (TD) is the total of the shortest distances from a node (n) to the other nodes in a system; i.e. TD(n) is the total of line n (or column n) in the distance matrix of a system.
d. Relative Asymmetry (RA) measures accessibility of a particular space relative to adjacent spaces or degree of integration of a particular space to the building's structure. The resulting value varies between 0 and 1, with values approaching 1 indicating lower accessibility. RA = 2(MD − 1)/k − 2, where MD is the mean depth of the system and k is the number of spaces in the system.

Limitations and Delimitations

The road grid has been considered as the only unobstructed open space network that contributes to connectivity, permeability, and integration. Internal water bodies, channels, and creeks too were not included into the open space-web. Only the surge action that is being considered as the close companion of permeability, connectivity, and integration measures of spatial configuration (Kraus & Lin, 2009). Because surge damage is one of the three contributors to overall damage, significantly low significance levels are expected from both the One-way Analysis of Variance (ANOVA) and Logistic regressions. Because there is a no reliable criteria to calculate percentage damage from each type of hurricane action (wind, wave, and surge), low significance levels can be expected. Even though it is out of the scope of this study, without adjusting the significance level for percentage-damage from surge, the relationship between axial connectivity and levels of surge damage should not be concluded as being statistically undependable. To minimize the threat to internal validity, only damage to the foundation and the overall structure was considered for the One-way ANOVA. It is believed that roof damage was caused mainly by wind action, and therefore excluded from the analysis. An important limitation of this study was narrowing the scope of the syntactical analysis to settlements with the least number of vacation homes or housing rentals built for tourism purposes in order to avoid real estate developments' associated housing design and settlement planning skewing the results.

Conclusions

An amalgamation of tactics certainly can unveil hidden constructs of the socio-cultural production of place as well as space. Some of these may not be completely hidden, but most are interpreted out of context including their operational meaning in architecture. This is an inherent condition of research, recovery, and mitigation of disaster and mass emergencies due to their magnitude and urgent nature. The research design strives to ascertain space accumulation of progressive settlements as well as the architectural makeup of building elements from the smallest building blocks of such settlements. The goal was to broaden the knowledge base of disaster resiliency of a particular intact spatial unit from planning of settlements to designing of individual dwellings. While syntactical analysis furnishes an integrated map of spatial connectivity among open space networks of the settlements under investigation, the research rubric revealed the structural patterns of elevation among residential units. However, analysis of elevation pertaining to permeability should not be investigated in settlements that show no or insignificant correlation between settlement connectivity and surge damage. In this sense, the object specific studies should only be carried out in settlements where spatial connectivity informs aspects of surge permeability with regard to ground elevation of individual properties. Therefore, the need arise out of necessity in assigning credibility to elevation characteristics of surge permeability that emphasized in the disaster literature for such surge-resilient houses. It may further validate the investigation by finding the degree to which those characteristics networked with the grid of open spaces are fundamentally accountable for controlling permeability.

References

Bafna, S. (2003). Space syntax: A brief introduction to its logic and analytical techniques. *Environment and Behavior, 35*(1), 17–29.

Bolivar Chamber of Commerce. (2010). *Land use map showing left and right halves of Bolivar peninsula.* Retrieved June 23, 2010 from http://www.bolivarchamber.org/InteractiveMap.aspx.

Daniels, A. P. (2010). Bolivar Peninsula. *Handbook of Texas Online*. Denton, TX: Texas State Historical Association. Retrieved June 11, 2010 from http://www.tshaonline.org/handbook/online/articles/rrb06.

Darjosanjoto, E. T. S. (2007). Permeability maps of residential settlements within the coastal area of Surabaya, Indonesia. In *Proceedings of the 6th International Space Syntax Symposium*. Istanbul, Turkey.

Doran, K. S., Plant, N. G., Stockdon, H. F., Sallenger, A. H., & Serafin, K. A. (2009). *Hurricane Ike: Observations and analysis of coastal change* (U.S. Geological Survey Open-File Report 2009-1061). Retrieved January 11, 2011 from http://pubs.usgs.gov/of/2009/1061/pdf/ofr2009-1061.pdf.

Downs, R. M. (1970). Geographc space perception. In C. Board (Ed.), *Progress in Geography* (Vol. 2, pp. 65–108). London, UK: E. Arnold.

Food and Agriculture Organization of the United Nations. (2009). *FAO and traditional knowledge: The linkages with sustainability, food security, and climate change impacts*. Retrieved June 17, 2010 from ftp://fao.org/docrep/fao/011/i0841e/i0841e00.pdf.

Goldberger, P. (2009). *Why architecture matters*. New Haven, CT: Yale University Press.

Hall, E. T. (1966). *The hidden dimension*. New York, NY: Doubleday.

Hillier, B. (1996). *Space is the machine: A configurational theory of architecture*. Cambridge, MA: Cambridge University Press.

Hillier, B. (2004). Can streets be made safe? *Urban Design International, 9*(1), 31–45.

Hillier, B., & Hanson, J. (1984). *The social logic of space*. Cambridge, MA: Cambridge University Press.

Hillier, W. R. G., Hanson, J., & Peponis, J. (1987). Syntactic analysis of settlements. *Architecture et Comportement/Architecture and Behavior, 3*(3), 217–223. Retrieved June 21, 2011 from http://discovery.ucl.ac.uk/86/1/hillier-etal-1987-synactic-analysis-settlements.pdf.

Hillier, B., Penn, A., Hanson, J., Grajewski, T., & Xu, J. (1993). Natural movement: Or, configuration and attraction in urban pedestrian movement. *Environment and Planning B: Planning and Design, 20*(1), 29–66. https://doi.org/10.1068/b200029.

Institute for Business and Home Safety. (2009). *Hurricane Ike: Nature's force vs. structural strength*. Retrieved September 19, 2010 from http://www.ibhs.org/content/data/file/hurricane_ike.pdf.

International Code Council. (2011). *Codes, standards & Guidelines*. Retrieved May 2011, from http://www.iccsafe.org/cs/Pages/default.aspx.

Kraus, C. N., & Lin, L. (2009). Hurricane Ike along the upper Texas coast: An introduction. *Shore & Beach, 77*(2), 3–8.

Lukermann, F. (1964). Geography as a formal intellectual discipline and the way in which it contributes to human knowledge. *The Canadian Geographer, 8*(4), 167–172.

Markus, T. A. (1993). *Buildings and power: Freedom and control in the origin of modern building types*. London, UK: Routledge.

May, J. A. (1970). Kant's concept of geography and its relation to recent geographical thought. *University of Toronto Department of Geography Research Publications*. Toronto: University of Toronto Press.

Morrow, G. S. (2009). Analyzing the invisible: An assessment of the applicability of space syntax analysis to ritual and domestic architecture at ancient Tiwanaku, Bolivia. *Texas A&M Library Digital Repository* (UMI MR61632). Retrieved July 20, 2010 from ProQuest Digital Dissertations Database.

Rapoport, A. (1990). System of activities and system of settings. In S. Kent (Ed.), *Domestic architecture and the use of space* (pp. 9–20). Cambridge, MA: Cambridge University Press.

Tuan, Y. F. (2001). *Space and place: The perspective of experience*. London: University of Minnesota Press.

United States Census Bureau. (2003). *Texas: 2000, population and housing unit counts*. Retrieved May 9, 2010 from http://www.census.gov/prod/cen2000/phc-3-45.pdf.

Upton, D. (1983). The power of things: Recent studies in American vernacular architecture. *American Quarterly, 35*(3), 262–279.

Venturi, R. (1977). *Complexity and contradiction in architecture*. New York, NY: The Museum of Modern Art.

Chapter 5
Connectivity: Gaps and Overlaps

> Don't architects have a responsibility to make the world better, as Vitruvius and Alberti would remind us? Does an ugly public housing project that provides a home for fifty families not serve a larger purpose than a more attractive one that gives only twenty families a roof over their heads? Wasn't it the job of architects to try and solve the problem of rebuilding New Orleans after Katrina? (Goldberger, 2009, p. 35)

Even though this study is a spatial investigation that fundamentally focuses on the syntactical analysis of the Bolivar open space system, the analytical tool Space Syntax only furnishes connectivity measures for axial segments in the Bolivar neighborhoods. However, both DepthMap and Space Syntax are considered best for settlement analysis as an authoritative medium by means of graphical representation of connectivity among the open space network in the settlements under investigation. The results and discussion accompany an introduction to the nature of the damage following Hurricane Ike, along with a detailed description on the cohort of settlements and their axial dynamics. The physical attributes of the Bolivar settlements are presented from the point of view of axial integration and the nature of the damage. An inevitable comparison on nature of surge damage is also made to Bolivar Peninsula's neighbor, Galveston Island. In addition, the visibility maps demonstrate different levels of physical connectivity among linear developments of the gridiron patterned artery and associated open space network in different parts of the Peninsula. In essence, the twofold data analysis employs a study of hurricane damage of the settlement at the neighborhood scale and of individual dwelling units at the axial line scale. The use of syntactical measures to interpret spatial transactions between physical connectivity and various levels of hurricane damage forms the major portion of the chapter. This is followed by an establishment of the statistical method and tools and the theoretical underpinnings of both an Analysis Of Variance and a Logistic Regression to unveil the possible relationships hypothesized in the research questions. The final segment of the results and discussions analyzes damage to individual properties based on their connectivity and the damage index. Residential properties along most connected open space network in one of the highly connected Bolivar neighborhoods at Port Bolivar have been investigated using a logistical regression on data obtained through a permeability rubric. Logistic Regression is the statistical tool

C. Subasinghe, *Rebuilding for Resilience: A Barrier Island Case*,
https://doi.org/10.1007/978-3-030-65532-7_5

used to interpret significance portrayed in the rubric, based on availability of binary data (*no damage* or *destroyed*) and on the categorical nature of the permeability characteristics.

Bolivar Neighborhoods: The Nature of Damage

Typically, wooden framed single-family residences elevated on wooden logs or pilings are the most common housing style found on both Bolivar Peninsula and west Galveston Island. Most of the vacation homes located directly along the shoreline employ more sophisticated mainstream building methods and finishes than landward houses. Generally, elevation of floor platforms varies from two to four meters above grade or four to six meters above mean sea level. Figure 5.1 illustrates predicted inundation using modeled storm surge based on Hurricane Ike storm conditions and landfall location. Actual wave-surge directions and surge damage variations along the Gulf of Mexico are shown in Figs. 5.2 and 5.3. Figure 5.3 further demonstrates

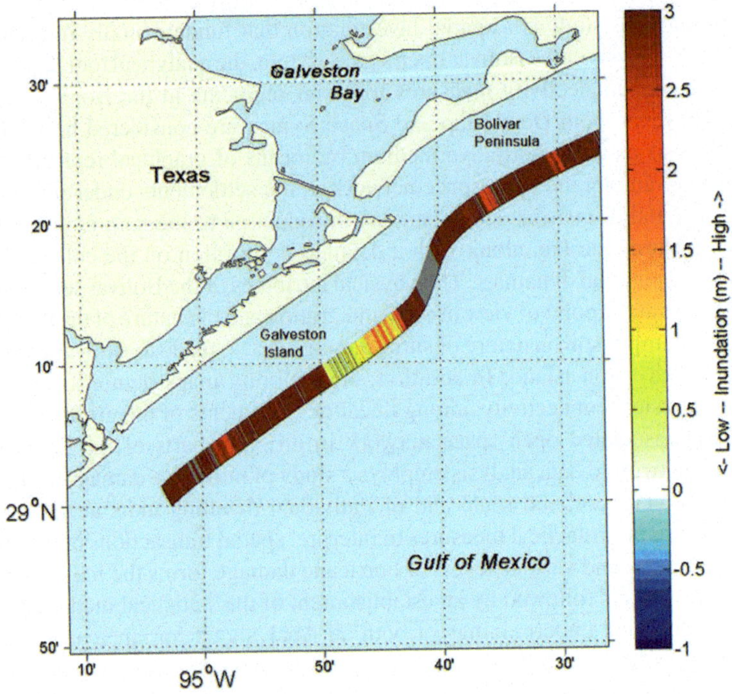

Fig. 5.1 Assessment of inundation potential (storm surge minus dune elevation) for Galveston Island and Bolivar Peninsula (*Source* USGS, 2008)

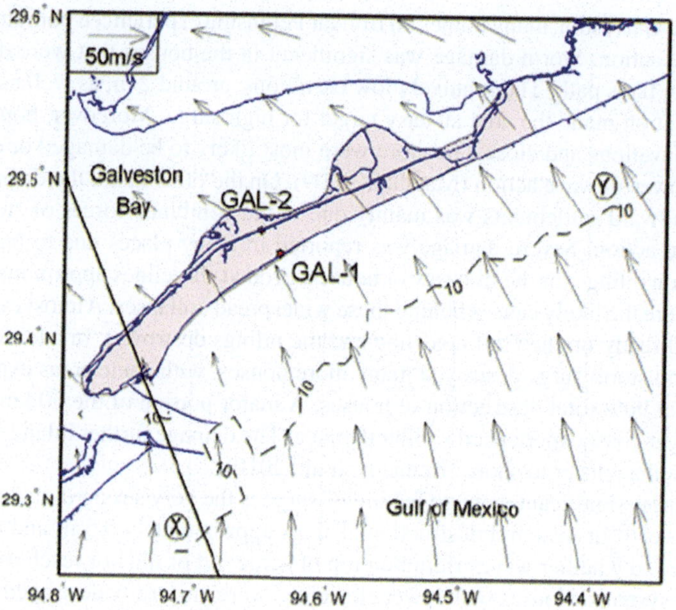

Fig. 5.2 Wind field and wave/surge directions during Hurricane Ike on Bolivar Peninsula (*Note* The vector wind field is shown at landfall and bathymetry contours in dashed lines. *Source* Kennedy et al., 2010)

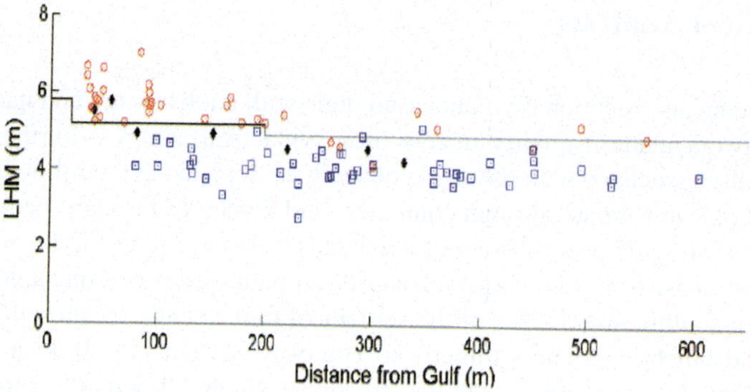

Fig. 5.3 Survival and destruction rates based on elevation and distance from the Gulf of Mexico shoreline (*Note* (□) Surviving houses. (□) Destroyed houses. (◆) Surviving, but significant wave damage. *Source* Kennedy et al., 2010)

survival and destruction against increasing distance from the Gulf of Mexico shoreline. The X-axis plots the Lowest Horizontal Structural Member (LHM) elevation and the solid line approximates 1993 FEMA Base Flood Elevation (BFE) with distance inland.

Unlike Galveston, many homes on Bolivar Peninsula experienced both storm surge and wave action. Storm damage was significant in the houses that were directly in Hurricane Ike's path. The relatively low elevations, around 2 m, as well as the open coast location made the area an easy target for high surge. Moreover, houses at the lowest elevations and closest to shore were most likely to be damaged or destroyed by the powerful wave action (Marshall, 2009). On the other hand, the damage to the more landward settlements was mainly due to the combined forces of storm surge and wave action. Severe damage was reported in many places due to undermined foundation pilings for the first row of houses. Eroded shorelines approximately 50 m inland were the likely causes behind these widespread collapses. Almost every house located directly on the Gulf coastline had the pilings destroyed. In addition, strong waves and storm surge destroyed many more houses, with some areas experiencing more than 90% total destruction of houses. A major portion of the 30 km length of the Bolivar coast experienced similar damage. The damage further inland was not as severe as that closer to shore (Kennedy et al., 2010).

The most significant damage feature of surge is the very strong division in elevation between survival and destruction. This is approximately 0.5 m and appears to be related to whether waves riding on top of surge did or did not reach the elevated flooring systems. Houses with lower elevations were uniformly destroyed by waves, while houses at higher elevations survived. The sheer weight and force of debris swept away by the floodwaters caused damage to properties beyond the shoreline.

Syntactical Analysis

Bolivar being an exceptionally flat coastal dune with unobstructed and narrow open space network made it qualify to rely on a syntactical analysis. Furthermore, its permeability associated with physical connectivity depicted in the following dataset had potentials to validate through computer fluid dynamics through secondary data available with USGS. The system of axial and convex space of the Bolivar coastal settlements consists mostly of a series of gridiron-patterned streets that link the northwestern or southwestern edges of the settlement to the southeast shoreline. Crystal Beach and Gilchrist connect directly to Highway 87, but Port Bolivar and High Island connect to the highway via the Broadway Avenue, Loop 108, and Highway 124, respectively. The Rollover Pass across Rollover Bay connects Bolivar Peninsula to the mainland, and Highway 87 connects to Galveston Island from the southeastern side.

Syntactical representation produces two forms of extension of open spaces for these settlements. The first form is a series of built spaces defined as a one-dimensional line that represents the connectivity of adjoining blocks. Hence, the fundamental argument behind this axial line representation is that the network of open spaces serves as a circulation route within the settlement. The second form does not necessarily connect open spaces to each other with a common property line; it is defined by boundary demarcations such as walls, fences, or earth embankments. This demonstrates qualities of an independent spatial compound. Combined together, the axial and convex representations provide an objective evaluation of spatial patterns of open spaces.

Axial Integration and Connectivity of the Settlements

As shown in the axial maps (see Fig. 5.4), both Port Bolivar and High Island are isolated from the main residential zone along the southwestern edge of the shoreline.

Crystal Beach and Gilchrist are part of the linear belt of neighborhoods facing the Gulf of Mexico. All other coastal settlements except High Island are linear in form and very regular in their patterns of axial space. Parallel streets, linked one to another, make Port Bolivar more compact than other coastal settlements (see Fig. 5.5). Axial maps transform the connectivity among open space network into a graph, which shows the relation between one axial segment and another.

According to Hillier and Hanson (1984), the most integrated areas ideally should carry the most movement. Among Bolivar settlements, the most integrated line of all axial lines is Highway 87, which runs across the Bolivar Peninsula connecting Galveston with High Island and southern Chambers County on the mainland. This also defines the northeast and southwest axes of the settlements and the access-spine for all settlements. From Gilchrist toward High Island, either Highway 87 or the Gulf

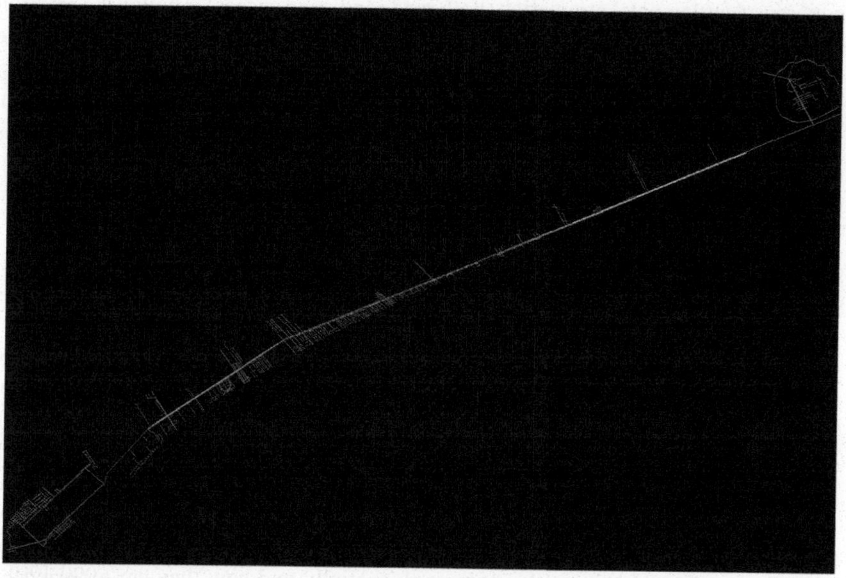

Fig. 5.4 Syntactical representation of the open space system of Bolivar Peninsula

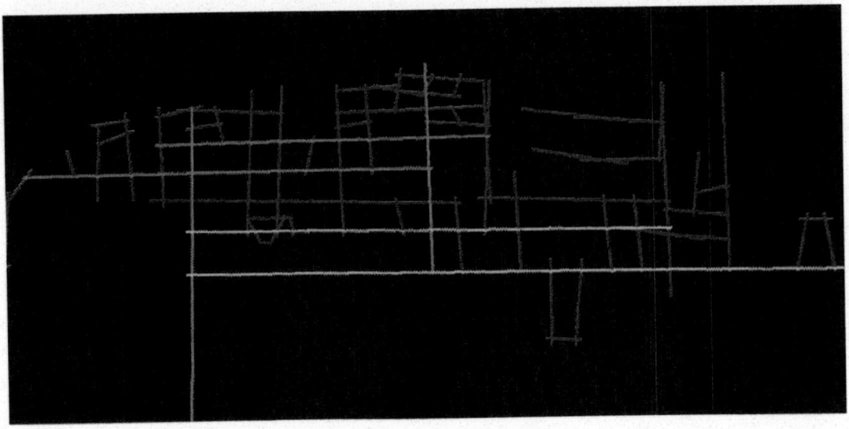

Fig. 5.5 Axial arrangement of the Port Bolivar settlements

Intercoastal Highway almost defines the northeastern shoreline, but it runs through the center of the settlements at Port Bolivar on the Galveston end. This is an important route because Point Bolivar, the western tip of Bolivar Peninsula, and Galveston are only separated by 2.7 nautical miles across Galveston Bay. However, the Crystal Beach, Caplen, and Gilchrist segment of the intercoastal highway (represented in a combination of red and yellow lines) marks the heart of the Bolivar integration continuum.

These lines are clear representations of the openness index with regard to the patterns of settlements' circulation (both roads and open spaces). Though the gridiron circulation routes of settlements are quite open and accessible, some segments close to the shoreline are more open and connected. The main northeast and southwest intercoastal highway is not only a part of the land-to-coast connection (global street), but it is also the origin of all streets running in north and southeasterly directions (local streets).

Visibility Mapping

An overall image of the open-space structure of the Bolivar settlements can be obtained simply by visual inspection. To make the discussion more systematic and rigorous, a set of recently developed techniques was employed for the representation and analysis of spatial structure, i.e. a visibility graph analysis (VGA) (see Fig. 5.1). In this case, the external space surrounding each of the settlements was treated as a single, unbroken element, and the 'depth' of spaces within each settlement was calculated accordingly. The relation between visibility and permeability is critical to an understanding of how settlements work spatially and how they are experienced by their occupants. VGA can help clarify this relationship.

As explained in the previous chapter, DepthMap, an automated method that performs a visibility graph analysis, divides the visibility measures into two groups. One group consists of global measures, such as mean depth and point depth entropy, while the other deals with local measures such as clustering coefficients and control. Because this study focuses on local measures, clustering coefficients assist in analyzing small networks. This is also useful in exploring the direction of junction points on the plan of a settlement or network of settlements. Even though the control does not directly contribute to the methodology of this research, it reports on relationships between a specific neighborhood and those immediately adjoining it. A number of properties inherent to the Bolivar visibility graph provide useful links to way finding and movement within a house or other building. As demonstrated in Fig. 5.1, the system of axial and convex space, low mean depth corresponds to a regular and linear layout, which facilitates movement in and out of the settlement. In terms of accessibility and visibility, clusters further away from Intercoastal Highway 87 require slightly more effort to get from one place to another. This may be due to the irregularity of shape compared to other clusters and the greater incidence of junctions and turning points within the spatial structure leading to open-space clusters.

Inferential Analysis (ANOVA)

The results of the ANOVA can be used to infer whether the means of the corresponding sample distributions also differ. Unlike t-tests that compare only two sample distributions, ANOVA is capable of comparing many in the Bolivar sample. Theoretically, it is possible to do multiple t-tests instead of a single ANOVA, but in practice, it may not be the appropriate strategy to obtain mean differences. On the other hand, ANOVA offers a number of advantages that t-tests cannot provide. For instance, ANOVA is robust in allowing the user to avoid the critical "inflated alpha" issue that may arise from conducting multiple t-tests.

Because this study seeks a relationship between connectivity and five different levels of hurricane damage, a one-way Analysis of Variance (ANOVA) test allows us to determine if one given variable has a significant effect on the second variable across any of the groups under study. A significant p-value resulting from a one-way ANOVA test would indicate that hurricane damage is differentially expressed in at least one of the groups analyzed. Because the surge damage has more than two groups being analyzed, the one-way ANOVA does not specifically indicate which pair of groups exhibit statistical differences. Therefore, a post hoc test was used in this specific situation to determine which specific pair/pairs are differentially expressed.

While n represents the number of items in each data set, the mean is calculated by summing all the individual items of data and dividing by the number of items of data (= sigma x divided by n). The variance or the measure of the spread of the data about the mean is calculated as follows:

$$\text{Variance} = s^2 = \hat{\sigma}^2 = \frac{\Sigma\left(x_i - \hat{x}\right)^2}{(n-1)}$$

The *F*-score and *p*-value of a one-way ANOVA will indicate whether the main effect of the independent variable "damage assessment" was significant. In other words, a significant F-statistic would tell us that class grade had a significant effect on average connectivity of the Bolivar settlements.

Informed Guesses and Extrapolation

The hypothesis for each damage group is there is no difference in the mean damage expression intensities in the groups tested. In other words, the damage has equal means across every group. Therefore, there is no difference in the mean damage expression intensities for average connectivity across all damage levels (ranging from *no damage* to *destroyed*) in the Bolivar neighborhoods. Table 5.1 illustrates the statistical representation of both the null hypothesis and the alternate hypothesis, based on the assumptions that the values in each of the groups (as a whole) follow the normal curve, with possibly different population averages and equal population standard deviations. However, the null hypothesis is that all of the group averages are equal.

Outcome Analysis

Table 5.2 shows the one-way analysis of variance (ANOVA) for average connectivity among five damage categories, using total sample of 248 houses at the α level of .05. As evident in the table, the highest sample size of 66 and 158 exists at extreme ends of the damage spectrum: *no damage* or *none* and *destroyed,* respectively. The smallest sample size (4) is for *major damage*, which means skewed results can be expected in the variances analysis.

The one-way, between-damage levels analysis of variance reveals a reliable effect of other level on physical connectivity, $F(4, 243) = 3.037, p = .018, MS_{error} = 15.796$, $\alpha = .05$. As shown in Table 5.3, the *p*-value of .018 suggests that a significant difference exists within comparisons of post-hurricane assessment in five damage levels. Because the *p*-value is less than the type I error or α level .05, the null hypothesis (all the means are equal) is rejected. Although the ANOVA results show that category-related differences exist in damage assessment, it does not tell us where those differences lie. Are the differences between *no damage* and *minor damage* or

Table 5.1 Statements of the null hypothesis and the alternate hypothesis

H_O	$\mu 1 = \mu 2 = \mu 3 = \mu 4 = \mu 5$	No relationship between mean physical connectivity of the Bolivar settlements and Hurricane Ike damage to houses
H_A	Not H_O	Some relationship between mean physical connectivity of the Bolivar settlements and Hurricane Ike damage to houses

Table 5.2 ANOVA descriptives for average connectivity

					95% confidence interval for mean			
	N	Mean	Std. deviation	Std. error	Lower bound	Upper bound	Minimum	Maximum
None	66	5.3106	4.36839	.53771	4.2367	6.3845	.00	20.00
Minor	13	3.0769	2.33494	.64759	1.6659	4.4879	.00	9.00
Moderate	7	4.5000	7.33144	2.77102	−2.2805	11.2805	.00	20.00
Major	4	.5000	1.00000	.50000	−1.0912	2.0912	.00	2.00
Destroyed	158	3.6693	3.74934	.29828	3.0801	4.2585	.00	28.00
Total	248	4.0474	4.03941	.25650	3.5422	4.5526	.00	28.00

$\alpha = .05$

Table 5.3 Average connectivity *F*-score and *p*-values for "between" and "within" groups

	Sum of squares	*df*	Mean square	*F*	Sig.
Between groups	191.917	4	47.979	3.037	.018
Within groups	3838.339	243	15.796		
Total	4030.256	247			

between *moderate damage* and *major damage*? Additional post hoc tests were done to address this issue.

The results of the Tukey HSD post hoc tests are shown in Table 5.4. The asterisks (*) indicate there is one pair of groups whose means differ significantly ($p < .05$) from each other. The hypothesis test indicated that the only visible relationship was between *none* and *destroyed*, at a .041 significance level and a 3.2422 confidence level. The *no damage* sample had the highest mean connectivity, and the reason for high standard errors for other groups maybe due to small sample sizes. The overall ANOVA showed a significant difference ($p = .018$), while the pair wise comparisons yielded only one difference (i.e., *none* vs. *destroyed*) that was strongly significant. Perhaps this is because the overall ANOVA compares all values simultaneously, thus weakening statistical power. On the other hand, the post hoc tests are simply a series of independent t-tests. The reason for high standard errors for other groups maybe due to small sample sizes.

As stated previously, the smallest sample size for *major* damage and the second smallest for *moderate* damage may be due to their close proximity to the *destroyed* and *no damage* or *none* categories. Perhaps the instrumentation error of the human cognitive process in assessing hurricane damage can explain such inconsistencies. Therefore, as presented in the means plots (see Fig. 5.6), connectivity and damage level significances for all damage groups except *no damage* or *none* and *destroyed* groups should not be misinterpreted as being similar sample sizes.

Table 5.4 Tukey HSD post hoc test

(I) Damage assessment	(J) Damage assessment	Mean difference (I-J)	Std. error	Sig.	95% Confidence interval	
					Lower bound	Upper bound
None	Minor	2.23368	1.20598	.346	−1.0808	5.5481
	Moderate	.81061	1.57983	.986	−3.5313	5.1526
	Major	4.81061	2.04652	.133	−.8140	10.4352
	Destroyed	1.64130*	.58249	.041	.0404	3.2422
Minor	None	−2.23368	1.20598	.346	−5.5481	1.0808
	Moderate	−1.42308	1.86321	.941	−6.5439	3.6977
	Major	2.57692	2.27243	.788	−3.6686	8.8224
	Destroyed	−.59238	1.14674	.986	−3.7441	2.5593
Moderate	None	−.81061	1.57983	.986	−5.1526	3.5313
	Minor	1.42308	1.86321	.941	−3.6977	6.5439
	Major	4.00000	2.49107	.495	−2.8464	10.8464
	Destroyed	.83070	1.53509	.983	−3.3883	5.0497
Major	None	−4.81061	2.04652	.133	−10.4352	.8140
	Minor	−2.57692	2.27243	.788	−8.8224	3.6686
	Moderate	−4.00000	2.49107	.495	−10.8464	2.8464
	Destroyed	−3.16930	2.01218	.515	−8.6995	2.3609
Destroyed	None	−1.64130*	.58249	.041	−3.2422	−.0404
	Minor	.59238	1.14674	.986	−2.5593	3.7441
	Moderate	−.83070	1.53509	.983	−5.0497	3.3883
	Major	3.16930	2.01218	.515	−2.3609	8.6995

*The mean difference is significant at the .05 level

Port Bolivar: Permeability Analysis

Based on the connectivity measures, Port Bolivar was selected for an in-depth analysis to reveal possible relationships between various physical attributes of individual properties that may or may not have had a direct impact on structural elevation in terms of permeability. As detailed in Chapter III, the permeability rubric was the culmination of elevation characteristics of surviving houses, which was reviewed in the literature as crucial in allowing surge to escape. Following the example of Hillier and Hanson (1984), convex size can be associated in this case with increasing connectivity to segments of space, rather than increasing permeability of buildings.

Settlement interface maps more or less convey the permeability of the settlement. The interface map (see Fig. 5.7) clarifies the fact that the system is shallow from outside the settlement. Some of the east-west open spaces represent convex spaces that are as long as axial spaces, allowing uninterrupted movement from the front of the dwelling to the shore. Thus, they not only provide the place for day-to-day

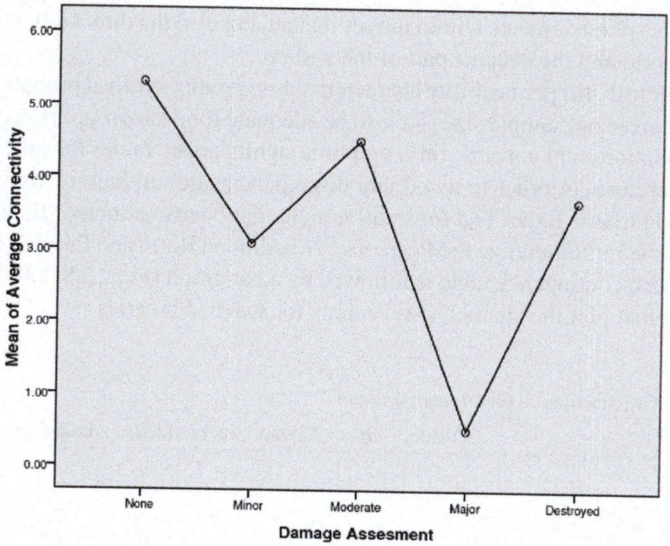

Fig. 5.6 Means plot for the one-way ANOVA

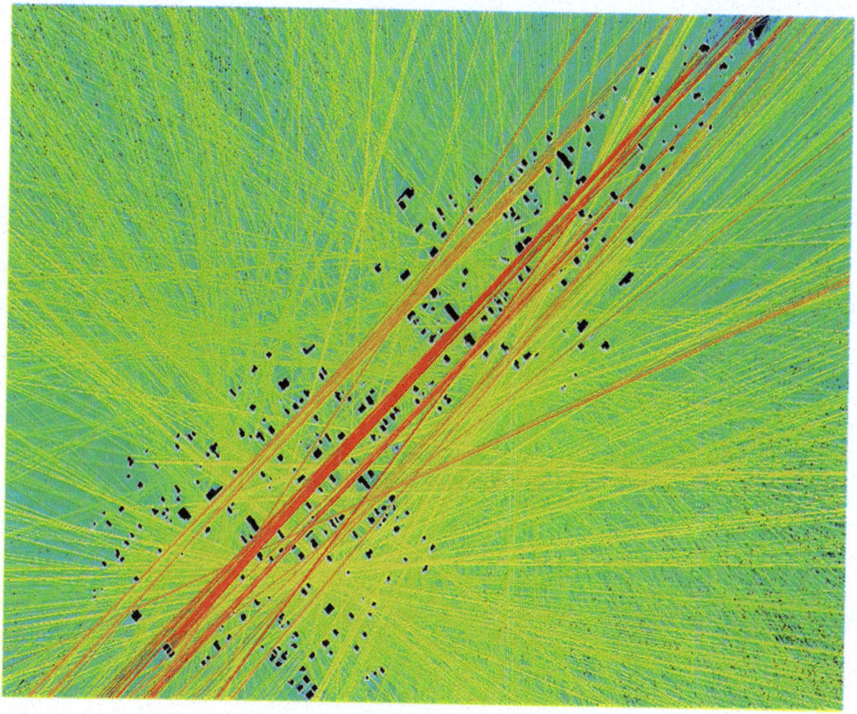

Fig. 5.7 All lines settlement interface map for Port Bolivar neighborhoods

occurrences or movements within the settlement, but also the direct link between the outside world and the deepest part of the system.

In order to distill permeability characteristics to predict survival probabilities via a logistic regression, sample sizes have to be adequate for each group. However, 2×2 cross tabulations will not perform asymptotic significances values for smaller sample sizes. Therefore, in order to avoid any discrepancies due to lack of significance, a two-tailed Fisher's Exact Test for small samples has been conducted. Tables 5.5, 5.6, 5.7, 5.8, and 5.9 summarize the Pearson χ^2 Likelihood Ratio and Fisher's Exact Test results. Each contingency table is followed by a bar graph (Figs. 5.8, 5.9, 5.10, 5.11, and 5.12) that plot the characteristic counts for survival binaries.

Table 5.5 Crosstabulation for s*treet orientaion*

	Value	*df*	Asymp. Sig. (2-sided)	Exact Sig. (2-sided)
Pearson χ^2				
Likelihood Ratio				
Fisher's Exact Test				.7749
Linear-by-linear Association	77			
N of Valid Cases				

Table 5.6 Crosstabulation for *ground elevation*

	Value	*df*	Asymp. Sig. (2-sided)	Exact Sig. (2-sided)
Pearson χ^2	10.635[a] 10.669	3.3		
Likelihood Ratio			.014	.01175
Fisher's Exact Test	8.446	1		
Linear by linear Association	77		.004	
N of Valid Cases				

[a]One cell (12.5%) has an expected count of less than 5. The minimum expected count is 3.16

Table 5.7 Crosstabulation for *ground enclosure*

	Value	*df*	Asymp. Sig. (2-sided)	Exact Sig. (2-sided)
Pearson χ^2	14.301[a] 17.179	3.3	.003	
Likelihood Ratio			.001	.01456
Fisher's Exact Test	5.009	1		
Linear by linear Association	77		.024	
N of Valid Cases				

[a]3 cells (37.5%) have expected count less than 5. The minimum expected count is 2.10

Table 5.8 Crosstabulation for *external bracing*

	Value	*df*	Asymp. Sig. (2-sided)	Exact Sig. (2-sided)
Pearson χ^2				
Likelihood Ratio				.2057
Fisher's Exact Test				
Linear by linear Association	77			
N of Valid Cases				

Table 5.9 Crosstabulation for *external attachment*

	Value	*df*	Asymp. Sig. (2-sided)	Exact Sig. (2-sided)
Pearson χ^2	2.167[a]	3.3	.538	
Likelihood Ratio	2.111		.550	.5417
Fisher's Exact Test		1		
Linear by linear Association	.033		.836	
N of Valid Cases	77			

[a]3 cells (37.5%) have expected count less than 5. The minimum expected count is 1.05

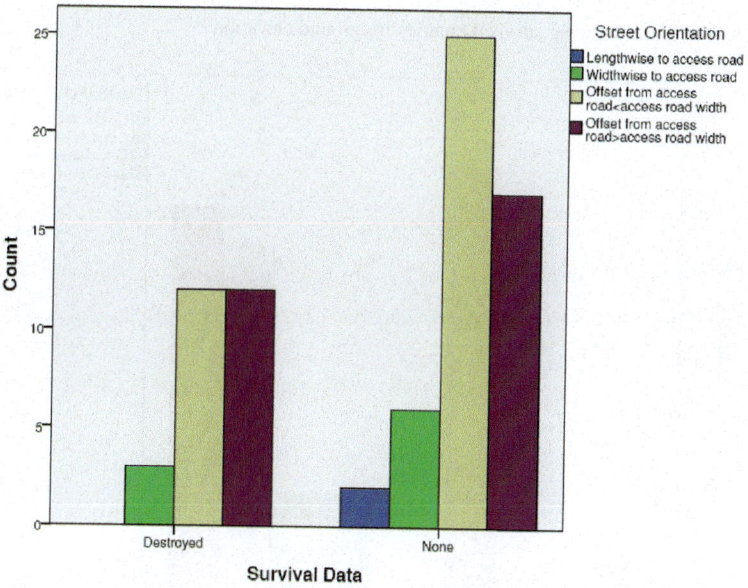

Fig. 5.8 Bar graph showing survival binaries for *street orientation*

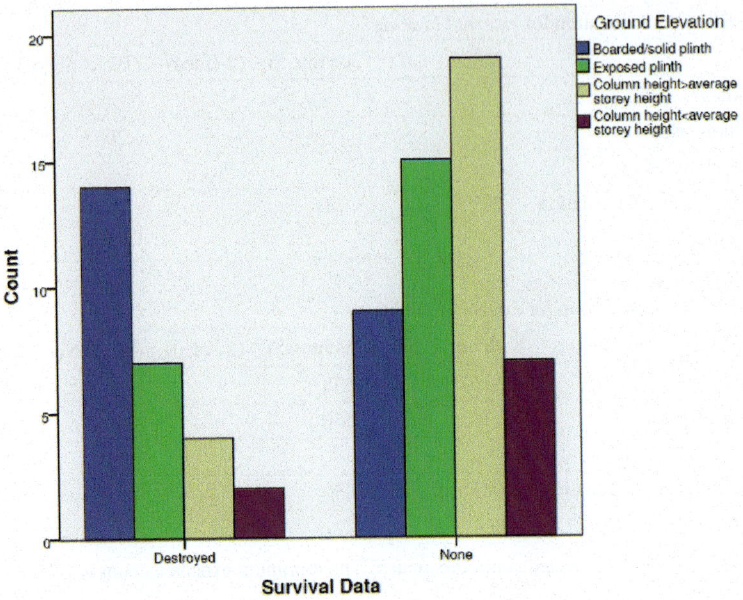

Fig. 5.9 Bar graph showing survival binaries for ground elevation

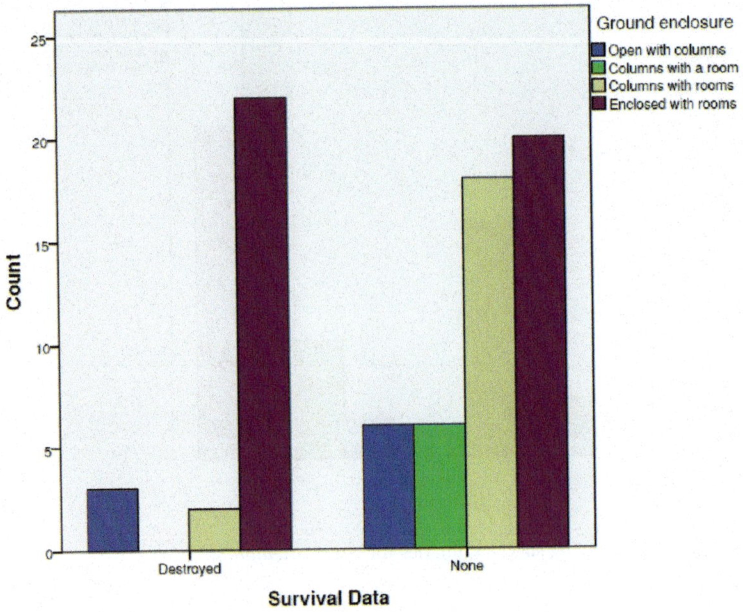

Fig. 5.10 Bar graph showing survival binaries for *ground enclosure*

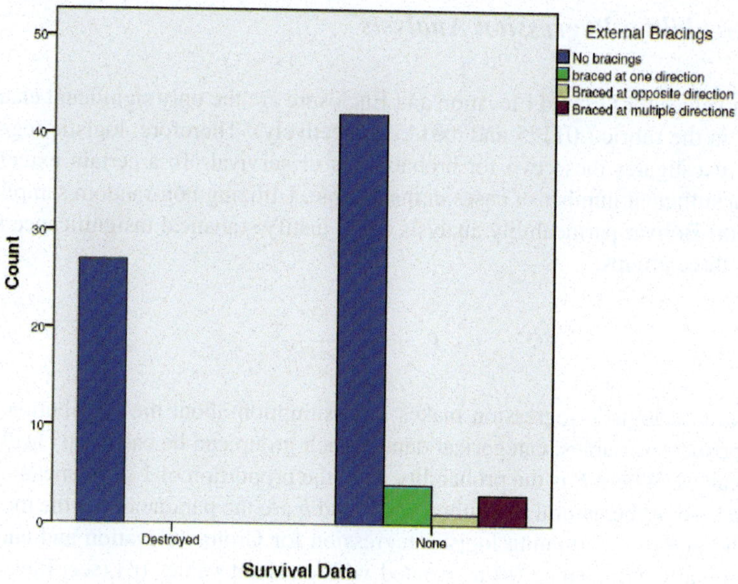

Fig. 5.11 Bar graph showing survival binaries for *external bracing*

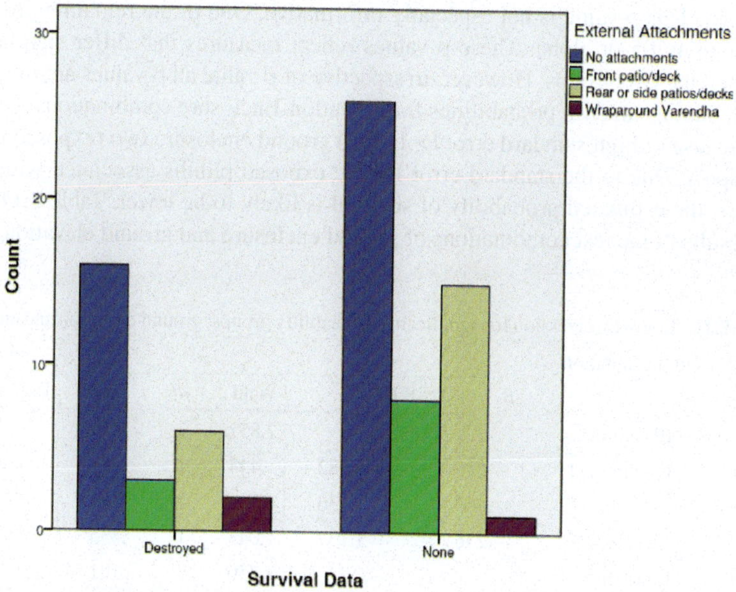

Fig. 5.12 Bar graph showing survival binaries for *external attachment*

Permeability: Regression Analysis

Exact values for Ground Elevation and Enclosure are the only significant characteristics in the rubric (.01175 and .00145 respectively). Therefore, logistic regression only investigates these two for probabilities of survival. To a certain extent, both lack a sufficient number of cases in the sample. Utilizing non-random sampling for the Port Bolivar permeability analysis could justify statistical insignificance for the other three groups.

$$P = \frac{e^{a+bX}}{1 + e^{a+bX}}$$

Because logistic regression makes no assumption about the distribution of the independent variables, categorical data in each group can be easily be checked for tendencies. Where *P* is the probability of 1 (the proportion of 1 s, the mean of Y), *e* is the base of the natural logarithm, and *a* and *b* are the parameters of the model.

The purpose of running logistic regression for Ground elevation and enclosure is to obtain significance (*p*) associated with constant values (β) (see Table 5.10). These p values tend to be conservative if the sample size is limited or the sample is not sizeable. In this instance, the p value associated with the constant exceeds .05 or alpha. This *p*-value is not especially informative. One of the remaining p values is less than .05 or alpha. These p values reflect measures that differ significantly between the two groups. However, irrespective of p value all β values are important in calculating survival probabilities for Elevation-Enclosure combinations.

Because of high standard error (> 16302) ground enclosure two (exposed plinth) is suspect. Due to the standard error for the exposed plinths associated with these houses, the estimated probability of survival is likely to be lower. Table 5.11 gives the results of the test combinations of ground enclosure and ground elevation group

Table 5.10 Logistic regression for significant permeability groups: ground elevation and enclosure

Variables in the equation

		B	S.E.	Wald	*df*	Sig.	Exp(d)
Step 1[a]	Elevation			2.852	3	.415	
β_1	Elevation (1)	−.635	1.082	.444	1	.557	.630
β_2	Elevation (2)	.463	1.136	.166	1	.683	1.589
β_3	Elevation(3)	−.236	1.137	.043	1	.835	.789
	Enclosure			4.750	3	.191	
β_4	Enclosure (1)	.933	.958	.950	1	.330	2.543
β_5	Enclosure(2)	21.140	16,302.252	.000	1	.999	1.517E9
β_6	Enclosure (3)	2.367	1.086	4.750	1	.029	10.662
β_0	Constant	.014	1.038	.000	1	.989	1.015

[a]Variable(s) entered on step 1: Elevation, Enclosure

Table 5.11 Logistic function of survival probabilities

		Ground enclosure			
		Enclosure[1]β_4	Enclosure[2]β_5	Enclosure[3]β_6	Enclosure[4]β_0
Ground elevation	Elevation[1]β_1	.624	41.038	3.492	−1.242
	Elevation[2]β_2	3.76	43.243	5.688	.954
	Elevation[3]β_3	1.422	41.836	4.29	−.444
	Elevation[4]β_0	1.894	42.308	4.762	.028

elements for survival probabilities. For the interpretation of logistic Table 5.11, the estimated probability of survival for a house with ground elevation one and ground enclosure one can be predicted as .62%. As apparent in the table, even though ground enclosure two is suspect, the highest probabilities for survival lie in the interactions of ground elevation and ground enclosure two.

Conclusions

Historically, Hurricane Ike is considered one of the most devastating hurricanes to have made landfall on Bolivar Peninsula. This catastrophic event created a more complex dynamic for hurricanes vs. settlements. First, it oversimplified the magnitude of the destruction to built environments in terms of the massive numbers. Second, it brought to light the hidden or unknown spatial underpinnings of transaction between natural events and built fabric at both the settlement and dwelling levels. Simple fluid dynamics authorizes behavior of surge flow from high to low pressure zones. Depth, connectivity, and the layout pattern of open spaces alter surge flow. Built mass, ground elevation, and structural integrity of the properties manipulate surge flow.

A nexus between connectivity of open space network and hurricane damage from surge does exist; flood escape routes do get affected by built masses and specific spatial and built characteristics of settlements and houses, which are their building blocks. Typical gridiron patterned road arteries established by early Spanish settlers still plays a significant role in the intercoastal transportation system and in the local and global movement systems of Bolivar Peninsula. While *no damage* and *destroyed* damage levels directly deal with settlement syntactic analysis, the *minor*, *moderate*, and *major* damage levels suffer from small sample sizes. The only significant groups in the permeability rubric, ground elevation and enclosure, also exhibit a similar statistical relation to survival binaries. The high standard error for ground enclosure two, the exposed plinth (β_5), makes statistical significance for its combinations with ground enclosure suspect. However, given the high number of variations between the results used to calculate the mean value for the Broadway Avenue sample (due to the small sample size for ground enclosure two), the mutual impact between a group and a group characteristic cannot simply be ignored.

References

Goldberger, P. (2009). *Why architecture matters*. New Haven, CT: Yale University Press.

Hillier, B., & Hanson, J. (1984). *The social logic of space*. Cambridge, MA: Cambridge University Press.

Kennedy, A., Dosa, M., Zarama, F., Gravois, U., Zachry, B., Rogers, S., & Sallenger, A. (2010). Inundation and destruction on the Bolivar Peninsula during Hurricane Ike. In *Proceedings of the International Conference on Coastal Engineering No 32*. Shanghai, China: Coastal Engineering Research Council. Retrieved July 10, 2011 from http://journals.tdl.org/ICCE/article/view/1379/pdf_338.

Marshall, T. P. (2009). *Hurricane Ike damage survey*. Poster presented at the 29th Conference on Hurricanes and Tropical Meteorology, Tucson, AZ. Retrieved March 2, 2021, from https://ams.confex.com/ams/29Hurricanes/webprogram/29HURRICANES.html.

United States Geological Survey. (2008). *Coastal change hazards: Hurricanes and extreme storms*. Retrieved May 21, 2010 from http://coastal.er.usgs.gov/hurricanes/gustav/coastal-change/.

Chapter 6
Rebuilding Rhetoric Aka Significance

> There can be no vulnerability without risk; there can be no community without vulnerability; there can be no peace, and ultimately no life, without community. (Peck, 1998, p. 233)

To a considerable extent, people do know how to build their abodes to sustain not only their socio-cultural environment, but also to make the best out of the climatic conditions in places where they live. However, they tend to compromise in planning, designing, and constructing their settlements and abodes due to both economic and ecological reasons. Economic reasons are more obvious than non-economic reasons such as the euphoria of being integrated or accepted into a mainstream planning and building precinct. What is enigmatic about the spatial-aspatial dialectic among different abodes in spontaneous settlements is may be their dynamic dealings with extreme natural events. Hence, an objective evaluation of such responses could be conducive to the processes of grounding complex concepts pertaining to abstract yet experiential idioms of spontaneous settlements as well as their responses to adverse environmental reactions. Thus, the degree of precision does not depend solely on the processes and procedures, but also on the basic assumptions by which such processes and procedures are assembled. Consequently, the conclusions presented in this discourse should be looked at as an antidote for a phenomenon which is ever changing, and which offers a challenging set of diverse opportunities in forms of critical constrains. These opportunities manifested in constrains oscillates between dependency on external support and resiliency through capacity building or the ways in which stakeholders manage disaster vulnerability and sustainability by calibrating risks against recovery (Fig. 6.1).

Conclusions for Policy

Not different from any other mechanism of human survival, disaster coping tactics are the result of a long-term give and take process. On the other hand, post disaster recovery has always been presented as a conglomerate of financial burdens on

C. Subasinghe, *Rebuilding for Resilience: A Barrier Island Case*,
https://doi.org/10.1007/978-3-030-65532-7_6

Fig. 6.1 Disaster vulnerability vs sustainability model based on risk deduction and recovery optimization

displaced people, taxpayers, non-profit organizations, and state and federal governments rather than a wakeup call for rethinking development and revisiting sustainability toward resiliency. One reason for this may be that development in disaster-prone areas and a tacit knowledge of spontaneous settlements has not been looked upon as something that can be analyzed in a sustainability framework. While mortality rates and economic losses are at the forefront of disaster statistics, recovery funding is the devil's advocate. Assigning a monetary value to economic gain from tacit knowledge in the rebuilding stage of the recovery process is not as straightforward as allocating value for market-based standardized methods used in conventional rebuilding systems. Nevertheless, the question is whether to let people to continue to live in disaster-prone, environmentally sensitive areas, or to turn disaster damage into something constructive and truncate or control the growth of such settlements through planned relocations and resettlements. Also, to be noted is, regardless of their intuitive tacit wisdom, locals do not always prepare for the worst-case scenario.

In the broadest sense, this discourse should not be taken as an excuse to continue exploiting hazard prone locations that are intended ecologically for natural disaster absorption or mitigation purposes. The development and growth of barrier islands such as Bolivar Peninsula and Galveston are a prime example of repercussions from oversighted spatial planning and overlooked spatial practices. Such planning and practices spontaneously produce substantial spatial agglomeration along with all the implications associated with such development patterns. Therefore, the strategic replacement of identified built parcels that can be severely affected by natural hazards with spongy open areas for disaster absorption should be imperative to any post disaster recovery agenda. This theory could be applied to remote or suburban areas with urban characteristics similar to Bolivar Peninsula. The acknowledgement and recognition of human settlements that have survived decades of natural hazards because of their unique building practices and technical wisdom should not be used as a defense for urbanization of sites vulnerable to natural hazards. Moreover, what is crucial about this phenomenon is its vulnerability to severe misinterpretations as an obstacle for human settlement growth. Even though almost every disaster study critically analyzes the causal relationships of the disaster-damage dichotomy, most fail to include specificities for revitalizing eco-sensitive coastal urban areas. In particular, they fail to take into account where protecting environmentally sensitive land parcels from further exploitation should stand in the post disaster rebuilding process.

Post-disaster Housing on Bolivar Peninsula

FEMA endured numerous critiques and disparate discourse on federal (government) lethargy over Hurricane Katrina post-disaster management. However, in the wake of the Hurricane Ike catastrophe, FEMA appropriately disseminated four different types of housing needs assistance. By looking at the following housing assistance schemes, one may easily conclude that different types of assistance programs tailored to address different levels of damage and housing needs are comprehensive enough to take the situation under control. First, *temporary housing* provided financial assistance for a specific, limited time so that people could rent an alternative place to live or in absence of suitable rental properties, a government provided housing unit. Second, FEMA disbursed funding for homeowners to *repair* their damaged primary residential units that were not covered under any type of insurance scheme. This category of assistance focused on properties under *minor* or *moderate* damage. In addition, it also addressed the immediate basic needs issues pertaining to making the damaged home safe, sanitary, and functional. The third type of assistance accommodated all other damage levels, including properties with *major* damage and *destroyed* properties; these were served under *replacement* assistance. Furthermore, FEMA provided funding to replace uninsured damaged homes with new ones, but the policy was not specific about whether or not the assistance facilitated full reconstruction costs for replacement housing. Fourth, the direct assistance or monies for *permanent housing construction* occurred only in insular areas or remote locations identified by FEMA where no other type of housing assistance was possible. However, three years after Hurricane Ike, many settlements still have as much as 30% of the housing blocks in need of either *major* repair or *replacement* housing.

Conclusions for Practice

Survival maybe contextual, but the Bolivar survival formula ascertained conclusive understanding of the potential for utilizing tacit knowledge in post disaster rebuilding. Results from the syntactical analysis revealed people's implicit awareness of where and how to build to reduce the impact from extreme natural events, as shown by traditional settlements in seismic-prone Turkey. However, the syntactical analysis revealed that, although the most of the survived houses were built in areas where the connectivity was high, this relationship was not conclusive for Bolivar settlements. Moreover, it is possible that gridiron access roads and associated open spaces established by early Spanish settlers of Bolivar Peninsula contributed to surge permeability among these settlements. From the urban planning perspective, this may support integrating open spaces to make an interconnected system for more uninterrupted escape routes and better options for surge permeability. Different levels of integration among different sizes of open space systems might even give different results for surge escape. Because they alter the intensity of wave action related surge damage, Bolivar Island has proven that open spaces adjacent to the shoreline are crucial not only for survival of the seaward settlements, but also for the landward settlements. The bulk of the debris load that intensified the surge force was created in such locations. Except for structural frames, the wall claddings of vernacular coastal settlements in many parts of the world are constructed with flimsy organic materials that can easily soak up the water from the surge. As a result, the debris load associated with the surge force becomes less or even insignificant. In contrast, the permeability analysis for Port Bolivar revealed minor significance for landward settlements in terms of debris load. Irrespective of its small sample size, the elevation characteristic with the highest potential for debris creation, external attachments, did not furnish significant data in the descriptive statistics for survival probabilities.

As confirmed in the research literature, local knowledge is not always effective if agents and agencies are not prepared for the consequences of the trial and error or piloting approach to rebuilding. Furthermore, the unpredictability of the severity of future disasters often leaves people with limited options, including readily available solutions from standard practices and existing building codes. In addition to increasing severity, increasing frequency of natural hazards also challenges the ability of local experts to recover with limited ways and means. In many situations, the additional time needed for local knowledge to adjust and settle for recovery process is not readily available. Contrarily, it takes considerable time for local knowledge and practice to accumulate and to be mobilized among the various people of an affected community. Notably, houses in Port Bolivar that are 70 years or older have demonstrated strong survival capacities because they only needed a few cosmetic repairs to external siding after severe surge attacks during Hurricane Ike. This is particularly valid for the cottage type houses on wooden logs a few feet high or the "Old and Gold" houses detailed in the following Semiotics of Survival section. The cottage type has proven to be the most surge resilient in landward settlements, but the houses elevated on wooden columns with no or a very few partitions are generally considered the most successful in allowing the surge to flow through. Similar houses in many parts of the world, including *marano* houses found in the flood prone villages of the Philippines, display more versatility whether they are located landward or seaward. In contrast to heavily intricate living compartments on the elevated upper floors or decks, the shear bulkiness of simple and crudely finished wooden logs or columns assures the least obtrusive ground spaces and structural stability.

Nexus: Mapping and Structured Observations

As evident in the demonstrated case, the properties from the areas where ample open spaces are integrated to provide movement and connections through an uninterrupted open space network were less vulnerable to surge associated hurricane damage. This relationship was particularly strong for survival extremes: *no damage* and *destroyed* categories. However, it was less significant for statistical interpretations for *minor*, *moderate*, and *major* damage levels. It is possible the sample sizes may be too small for these categories. Similar patterns of results were demonstrated in specific built characteristics pertaining to permeability of survived houses in the Port Bolivar neighborhoods. Out of five groups of surge permeability features, only two groups strongly demonstrated higher probability percentages for survival during surge attacks form hurricanes. All elements of the group *ground elevation* portrayed a direct association to statistically significant survival rates, but when combined with *ground enclosure*, only the characteristic *columns with a room* (β_5) showed considerable significance. However, the survival probability function of logistic regression furnished further insights into myths and truths about local knowledge. Perhaps, not all ways and means demonstrated in self-built or locally built structures are hurricane resilient or least crucial when rebuilding after hurricanes, but they help provide insights when deciding what not do in future post hurricane recovery processes. Because this discourse does not include cost benefit analysis, especially for financial gains and losses, even less significant built characteristics of the studied settlement should not be taken lightly. Once stripped of heritage clichés, esoteric underpinnings of spatial-aspatial dynamics are easy to interpret in economic terms than in aesthetic terms.

Different modes of Hurricane Ike damage assessment, including this inquiry, have shown similar but different degrees of significances regarding building elevation and surge permeability. Furthermore, recent hurricane history has indicated that the numerous storms in the past few years that approached various parts of the Gulf Coast had surge levels much higher than the estimated 1.0% probability per year (100-year return period) surge levels. The most devastating hurricanes of the last decade (Hurricanes Ike, Katrina, and Ivan) had surge levels that were well above the expected annual Base Flood Elevation (BFE). In light of these events, it is appropriate for FEMA's National Flood Insurance Program (NFIP) to structure incentives to encourage more people to build above the BFE by several feet and preferably above the 0.2% annual probability for surge level (500-year return period). When all of the homes in a subdivision or community are essentially built at or below a single elevation, the risk that the entire area will be destroyed by extreme storm surge is greater. The current efforts of the Army Corps of Engineers and FEMA to remap surge inundation height around U.S. coastlines may produce new requirements for both 1.0% and 0.2% annual probability for surge level. The results of this research may also guide locals and builders to install decks at an elevation above the BFE, unlike some of the existing constructions where the decks are at the lowest habitable space. The second highest survival probabilities for *ground elevation* and *enclosure* combinations were exhibited for *enclosure two*, in which the column heights were greater than average story height, which objectively support the recommendations of the federal agencies.

In some cases, residents had moved their homes up an additional eight feet; this helped the houses withstand the possible 21-inch surge plus wave height that came with Hurricane Ike successfully. This was the case for the majority of the elevated wooden coastal structures on Bolivar Peninsula, and the evidence shows that it only took a few feet of water above the BFE to wipe out all homes built at or slightly above the minimum BFE requirements. Structures on Galveston Island also exhibited a strong link between survival and destruction and building elevation, although it was not as striking as on Bolivar Peninsula. Areas further from the Gulf of Mexico also showed considerable significance in this link. While a considerable number of buildings uniformly survived at higher elevations, noticeably some buildings at low elevations also survived including some at-grade homes built just above sea level. Along the edge of the Gulf of Mexico, the surrounding ground level infrastructures (e.g., sheds) were destroyed and subsequently swept from the area. Near the Intracoastal Waterway, more than 1.2 miles from the open Gulf, there were some houses that survived even when completely inundated. Flimsy objects like picket fences and plastic netting also survived inundation.

Semiotics of Survival

Hurricane resiliency should not be oversimplified into a permeability rubric or dwelling typology. Conceivably, the complex and contradictory forces of human creations can cause havoc during situations such as hurricanes, which are beyond human conclusiveness and thus, control. Although often incomprehensible to scientific wisdom, certain versatile disaster responses of what are commonly identified as folk, vernacular, indigenous, traditional, or spontaneous settlements are often temporary or instantaneous responses that disappear along with the peak of the event itself, thus leaving no residues for measuring and recording. Therefore, researchers may only be able to analyze the enigmatic residues of such reactions and responses. Perhaps the possibility of missing the essence of transactions between hazards and the milieu of man-altered physical fabric is greater than capturing it. The premise here is the generally accepted notion that the easiest possible access to the highest amount of rigorous field data from a disaster-stricken area is directly proportional to the time elapsed after a disaster event. However, after a certain time, this relationship becomes either constant or inversely proportional (see Fig. 6.2).

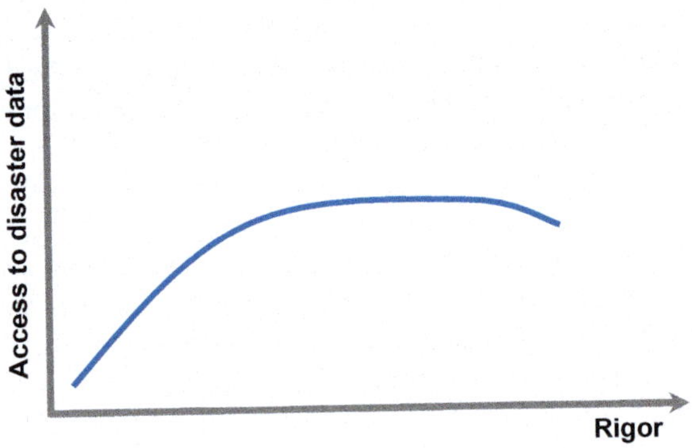

Fig. 6.2 After an indefinite time, the rigor of disaster data decreases as time passes and access to data increases

Obtaining accurate measures for in-between damage levels such as *minor*, *moderate*, and *major* is rather difficult because people tend to modify their houses either to make them habitable for further occupation or to qualify for insurance and FEMA assistance. Therefore, once the initial debris cleanup is done and the basic infrastructure is in place, the rigor and the access to data diminishes after an indefinite time. However, there a correlation exists between this indefinite post-disaster time lapse and the reestablishment of habitability among settlements.

The typologies presented here are not generic but are representative prototypes or models of pre-disaster dwellings that have undergone severe post-disaster ramifications. The permeability rubric facilitated distilling different permeability characteristics and their combinations in order to analyze representative abodes for varying resilience. An evaluation of the degree of each type of abode's resiliency characteristics is an amalgam of convex-axial in syntactical analysis and descriptive analysis in the permeability rubric. The representative names are reinterpreted versions from the colloquial identifications inhabitants used to differentiate houses of different ages, except for the "hill escape" type of house. The Hill Escape distinction has been seen among some real estate brochures that has been used to advertise its dominant multi-tiered roof formation, high elevation and flood proof feature.

Hill Escape

The "hill escape" type of house is the most contemporary response to protection against hurricane-surge attacks, as well as one of the most extreme responses. This approach alters the existing topography drastically by raising the ground elevation from 10 to 16 feet above existing ground level. Few houses have this capacity, and this approach was found the least in the entire Bolivar housing block. Inherent to this typology is erecting a typical suburban house without considering the prevailing environmental conditions or addressing such conditions at an arbitrary level. This is no better than an afterthought and is a result of incomplete design and planning often found in tract housing. Most of these houses are larger than the average house found in Bolivar neighborhoods and are relatively new constructions. Both the exterior and interior finishes display professional or hired labor skills than characteristics of self-built or locally built. Typically, these houses have an entrance lounge with wrapper-round verandas, three to five bedrooms, an open plan living-dining-kitchen arrangement, one to three garages, several restrooms, attic space, and many auxiliary rooms. Semi or full basements are not common in this category, although a very few had small underground storage units or wine cellars. Multi-tiered roof formations with several skylights or roof windows are also characteristic of this typology. Some of those roof cutouts are mere architectural features or "roof volumizers" and have no connection to interior spaces. In some cases, they are asymmetrically placed to maintain consistency of architectural façade, but still provide natural light to attic spaces. If they are operable, as in a few cases, they provide both light and ventilation. Because these houses are on artificially created hills, orientation is usually central

Fig. 6.3 A "hill escape" type of dwelling

and clearly visible from all four sides with a statement like presence. A ramp that runs to the main entrance creates a wide flat platform enabling convenient vehicular access and connecting the yard area in front of the main entrance with a garage. Some public buildings including the United State Postal Services building had more than one ramp to ensure an uninterrupted flow of vehicular traffic in and out of the area. Figure 6.3 illustrates a "hill escape" type of house.

Buddy System

The "buddy system" type of house is the second least prevalent, after the "high and dry" house. Two sheltering or dwelling units coexisting with different operational systems that are heavily dependent on hurricane season are characteristic of this type of house. Highly contrasting façade appearances as well as ground arrangement of the two units are striking features of this type. While the unit with the larger footprint usually accommodates the typical residential functions, the smaller unit usually provides storage facilities during the non-hurricane period of the year. A rectangular hurricane or refuge house is often oriented at a right angle to the main house. The interface between the two units is often demarcated by an open staircase that leads to the deck or small entrance/landing of the hurricane house. The ladder-like or in some instances spiral staircase of the smaller unit always has an uninterrupted physical and visual connection to the entrance of the main house; either it is in line with the main entrance or easily accessed without in-between spaces or barriers such as fences, vegetation, or different levels. Elevation of the smaller unit or rescue house usually exceeds typical story height. Ownership of this type of house is not very clear; it may belong to a single owner, several owners, or members of the extended family. The house's intended performance and operational procedures are only fully realized during hurricane season. During the rest of the year or season, these usually served as storage units. Some of the common features of the "buddy system" house include

Fig. 6.4 A "buddy system" type of dwelling

a sound structural system, simple spatial arrangement, easy access, and strategic orientation in relation to the main house. The exposed columned-ground floor of the hurricane/refuge unit sometimes houses fishing or leisure boats. Unlike the main house, the refuge unit is relatively uncrowded; vehicles parked in close proximity to these units were hardly noticed. Figure 6.4 illustrates the "buddy system" house.

Enough evidence existed to conclude that the immediate surroundings of the refuge/hurricane unit were intentionally kept as clutter-free as possible to ensure a smooth transferring of goods and items from the main house to the hurricane house in order to rescue them from hurricane damage. The main reason for keeping the hurricane house and its surroundings tidy is to be able to monitor threats to the structural integrity of the dwelling so that repairs can be completed without risk of structural failure.

Old and Gold

The "old and gold" dwelling is the most common type of residence along Broadway and Madison Avenues in the Port Bolivar neighborhoods. Most of these houses are more than 75 years old and are remaining pieces of typical South East Texas houses; they also are the smallest in square footage. With few exceptions, a great majority of these houses are single story buildings unspoiled by the lavishness of contemporary suburban living. These houses usually have a simple square plan layout, elevated on heavy hardwood logs two to four feet above ground, a wide front porch that runs throughout the front façade, an equally divided colonnade of four to six columns, and a steep hipped or gable roof. Many houses that belong to this category constituted separate structures dedicated to variety of functions including storage of fishing and yard maintenance equipment, covered parking, workshop facilities, and occasional secondary accommodations. Some houses may have wooden latticework covering the heavy timber thresholds that elevate the house above surge damage and various other climatic conditions. These lattices are primarily aesthetic in nature instead of serving as safety features, but they do prevent dust and debris from entering under the ground or first floor slab. They also make the underneath space less accessible

Fig. 6.5 An "old and gold" type of dwelling

to domestic animals and strays that could die in the in-between space or leave waste behind. The front façades of a few houses featured a false two-story appearance created by windows in the concealed area between the front porch-colonnade and gable end. Most of these windows are not operable and are not associated with a mezzanine or attic level. Typically, the main door is aligned with the entrance steps and a wooden gate that also makes up a small portion of the handrail of the front veranda. Generally, road frontage is wide open with no landscaping, keeping the unobstructed yard area between the access road and street front of the house. Usually they have a considerably bigger backyard that allows the smaller front yard to merge into the access road, creating a continuous stretch of open space along the road. Figure 6.5 depicts the "old and gold" house.

High and Dry

The "high and dry" type of dwelling is one of the most sensitive socio-cultural productions of space as a response to extreme local climatic conditions. It is also the least prevalent type of house on the Bolivar Peninsula. Even though a few of this type alter the existing ground level by creating an earth platform to receive first/ground floor columns, the artificial elevation is insignificant when compared with the "hill escape" houses. Island-like enclosed rooms among columns on the first/ground floor are characteristic of this type. However, first floor partitions are less structurally sound than those found on the second floor. These flimsy partitions are justifiable as they are created for temporary storage or occasional workshop facilities and are used primarily to provide shade from sun. Another key feature of these houses, column bracings or column anchors, make them more structurally intact. These are located approximately two to four feet below the level of the second-floor slab and

Fig. 6.6 A "high and dry" type of house

serve to prop and tie columns with the upper floor. Most of these bracings are on either opposite directions or two adjacent directions at corners; multi directional bracings were rarely seen. A freestanding staircase connecting to a wide upper floor deck creates a transitional space between the main entrance and staircase landing. More than one staircase and wraparound balconies make spacious additions to these houses. Similar to the "old and gold" houses, a wooden gate matching the handrail is always present at the top of the staircase. A recently constructed house had a roof opening similar to a sky light, roof window, or sun tunnel connected directly or via secondary space to a mezzanine or attic level.

Usually, the "high and dry" houses feature a paved ramp way that connects to any convenient and unobstructed entry to the first floor rather than connecting to the stairway to upper floors. Even though all the habitable places are on upper floors, this feature signals a strong presence of a ground floor that is not a merely used for storage but serves as a transitional space to both indoor and outdoor areas. The houses with wide wraparound verandas create a safety envelope around the habitable spaces of the upper floor while providing shade and additional covered areas for the ground floor. In addition, windows on the upper floor receive are shaded by the veranda roof overhang. The ensemble of veranda and overhang also acts as a cushion for strong winds. Figure 6.6 illustrates the "high and dry" house.

Modest Yet Mighty

Not only does the "modest yet mighty" type of house represent the most common dwelling type on Bolivar Peninsula, also it represents the majority of structures that survived Hurricane Ike. An elevated floor on freestanding columns roofed with a steep hipped or gable roof is distinctive to this type. This typology may be due to the accumulation of trial and error outcomes from cross-generational experimentation with hurricane resistance or survival at minimum socio-economic cost. Unlike other types, only a few variations of this type of house are seen in Bolivar neighborhoods. Even these few variations are restricted to a similar arrangement of upper floor balcony, overall dwelling size, direction, and orientation of staircases, and skylights or roof windows on the roof. The basic plan for this type of house is either a square or rectilinear in layout. The rectilinear houses are more common and often oriented lengthwise to the access road. They may have four different balcony arrangements: front, front, and rear, three-sided, or wraparound. Front balconies are most common, and windows can be seen in all walls in the wraparound scenario. Handrails and balustrades often seek support from ground to roof columns. In some cases, the column bracings or props that tie columns to the upper floor slab can only be seen among the peripheral columns. Because these houses lack partitions dividing rooms at the first-floor level, most of the ground/first floor has no paving or floor slab. In a few cases, only the foot of the staircase is paved to protect the wooden footings of stairs. Figure 6.7 depicts the "modest yet mighty" house.

The intermediate landing of the staircase can often be accessed from opposite directions. This creates a somewhat distinguished look as well as adding symmetry to the simple spatial organization. Even though most of the roof windows are not operable, they correlate to an attic or mezzanine floor arrangement. Most of the houses accompany separately built structures for workshop facilities, garages, or storage for fishing equipment. Typically, these are located to the side of the dwelling units rather than in the front or rear. Occasionally, attachment roofs were used as shelters for boats or RVs.

Fig. 6.7 A "modest yet mighty" type of dwelling

Significance: Rebuilding for Resilience

Perhaps, this discourse based on the latent characteristics of tacit knowledge related to sociocultural and econo-environmental production of space manifested in settlement layouts, building practices and technology concepts shed light into narrow aspects of resiliency. Such aspects might include people's desire to preserve their settlement identity, equitable social treatment, and an agreeable or justifiable timeline for rebuilding during the post-disaster recovery phase. Nonetheless, mitigating the effects of extreme natural disasters is difficult, especially when increasing natural hazards and decreasing direct funds for recovery leave limited options. This emphasized a holistic understanding of the complexity of disaster phenomena that enables effective assessment and wise disbursement of recovery assistance. The existing body of sustainable disaster recovery research found distinct faults with current rebuilding approaches, methods and materials and proposes that these be replaced with scientific research, streamlined technology, and approved building materials. Noticeably, a few of these research outcomes on the stated topics have ended up in good practices or building codes as revisions or addendums to existing regulations. In addition, the limited research on sustainable recovery either emphasizes the social aspects of recovery such as lack of community participation, dysfunctional agents and agencies, or methodological errors in implementation, or economic restrictions due to limited availability of aids and unorthodox distribution channels of assistance. Limited ethnographic work has been done regarding building practices via 'look closely at ground zero sites' to generate more grounded solutions. For example, spatial configurations of spontaneous settlements have never been a consideration for sustainability in majority of the studies in the field. Therefore, this discourse is intended to broaden the existing knowledge base by extending disaster inquiry into objective evaluations of the behavior of open spaces during hurricanes on settlement syntax and building practices. Three factors of significance in this discourse highlight the criticality of place character/identity, know-how justice, and speedy recovery for spontaneous coastal settlements in general. The findings of this study highlight three critical areas. First, there is a need to develop a system to evaluate spatial attributes of remote spontaneous settlements with subtle urban characteristics. Second, there is a need to assign objective values to the socio-cultural production of space and place. Third, there is a need to capitalize on local disaster resiliency strategies to concretize long-term sustainability toward resiliency.

Place Identity

In many theoretical and clinical circles, it was often questioned whether providing or enabling structurally sound hurricane resistant houses in places like Bolivar Peninsula would address the problem of hurricane survival. Most of the solutions take elaborate approaches that alienate physical context from socio-cultural variables.

Justification is necessary to avoid complicated value assessment of disaster-stricken settlements through an optimization of objective post-disaster rebuilding procedures. This discourse proposes and uses methods that assign operational values to products of human habitations that enable systematic and generalizable procedures for macro scale while acknowledging individually driven rebuilding efforts.

Aftermath of a natural disaster surface critical insights and thus, sizeable opportunities to preserve the inbuilt characteristics of the locality. Maintaining place identity for unhindered continuation of material cultures is strongly supported by different theoretical points of view and in numerous disciplines. Place identity can help to provide a sense of stability and continuity. Old buildings furnish evidence of history and permanence. People like to know not only where they are, but also "when," as well as how "now" relates to the past and the future. Space and time together are two of the major dimensions within which we live (Lang, 1994). A community-based approach to relief distribution and rehousing has not only proven successful, but also has greater potential to preserve the unique characteristics of the locality (Cuny, 1983). Rehousing that is based on local planning and skills while utilizing local knowledge and appropriate technology has the potential to preserve place character (Miculax & Schramm, 1998; Rawal & Desai, 1994). However, introducing alternate systems or modern conveniences to traditional settings that are completely alien to local character may amplify the disaster damage (Karenga, 1992; Williams, 1977).

Know-How Justice

Local knowledge or appropriated 'know how to do' has proven successful in levelling prevailing inequities among marginalized communities, which often become detrimental and catalyst to any rebuilding initiative aftermath of a disaster (Wisner, Gaillard, & Kelman, 2012). This resource-based disparities among different levels of social stratums tend to grow wider in disaster situations, particularly in absence of property insurances. They often comprise the most politically imperative as well as sustainably decisive part of the recovery effort, which usually demands immediate attention as well as critical intervention-often challenging the capacities of communities and collective of assistance. However, as also evident in this discourse, standardized solutions could fuel disparities instead of serving as social levelers. Acknowledging the building skills and practices of disaster victims and employing them in the post-disaster rebuilding process can potentially addresses an important part of the social justice with specific focus on integration of local know into rebuilding processes. Disaster stricken communities should not feel that they are passive recipients with no control over their post-disaster conditions. It is important that they be included as active participants in future growth asserting resilience. This notion is widely established in many academic spheres that not only strive to understand human instrumentation, but also seek ways to empower social equity in resiliency building process.

Direct benefits addressing social justice related to know-how integration have been captured in many literacy sources along with highlighting possible dire consequences of forsaking them for rapid recovery-the next significance of rebuilding for resilience. For example, many dimensions of mass disasters, including their causes and recovery processes, reflect profound ecological and social injustices catalyst to sustainable recovery (Ozerdem, 2003). Communities with significant social and economic disparity tend to be more vulnerable to the effects of disasters; indeed marginalized populations face greater challenges than do their privileged counterparts in terms of disaster recovery (Harrell & Michael, 2000; Kaiser Family Foundation, 2007; Morrow & Peacock, 1997). This suggestion certainly holds true in the case of Hurricane Katrina in New Orleans, a city besieged by environmental, social, and economic imbalances even before the landfall of the massive storm and subsequent levee failures of August 2005. By employing methods grounded in the insights of sustainable disaster recovery as well as critical discourse analysis, practitioners and scholars can uncover the political, economic, and cultural hegemonies that perpetuate injustice (Kincheloe & McLaren, 2000; Smith & Wenger, 2006). This discourse revealed leveling of prevailing social disparities during the recovery process via recognizing and employing resiliency characteristics that may present in spontaneous settlements.

Speedy Recovery

The resiliency-sustainability spiral developed as the core constituent of the theoretical underpinning for this discourse strongly suggests that speedy recovery leading to the return to normalcy is essential to long term resiliency against recurrent natural hazards (see Fig. 6.8). Bilello (2014) identified this process via a current-future binary in three levels of advancement from mitigation to transformation via adaptation.

Resuming normalcy or adaptation to changed status quo (new normalcy) for sustainable recovery is a core conclusion of this discourse. Therefore, elevating coping to a higher degree is probably the single best way to achieve sustainability for marginalized communities. The time needed for a disaster-stricken community to return to normalcy is decisive to sustainable recovery. Funding, which is the most critical factor of post-disaster rehabilitation and reconstruction, is significantly supported by recoverable damage as well as the use of easily replaceable and locally available materials and skills. The less stress placed on required funding, the faster the recovery—provided the affected community is eager to reoccupy their settlement under available external support. However, external systems sometimes consume considerable time to mobilise and establish in an unfamiliar locality where basic infrastructures are not in place to deliver required outcome. In contrast, local systems are often ready to mobilize and are versatile enough to adapt altered conditions to their advantage. Furthermore, resuming their lives in their respective homes simultaneously contributes to reinstating a displaced community to their former livelihoods easy. Subsequently, this may lead to vital reestablishment of the local economy and

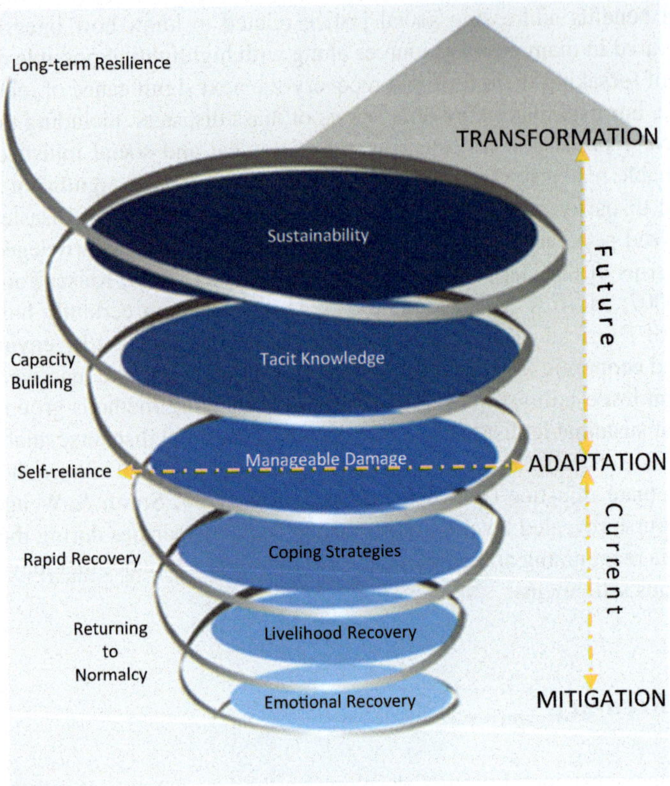

Fig. 6.8 Integrated sustainable disaster recovery model (Bilello, 2014) and Resiliency-sustainability spiral (*Source* Bilello, 2014 and Author)

thus ability to negotiate rebuilding programs with external agents and agencies on equal grounds. The research that supports speedy recovery as a mandatory rung on the recovery ladder also emphasizes the economic counterpart of the emotional recovery. Arguably, if no mortalities and major property damages involved, the strongest detriment to emotional recovery after a catastrophic event is livelihood security and reemployment possibilities. A settlement formed with units whose construction and maintenance costs are within the affordability limits of its occupants causes less financial burdens and the emotional recovery process is also relatively fast (Bolin, 1982; Bolin & Stanford, 1991).

References

Bilello, J. (2014). *Design for resilience: Mitigation, adaptation and transformative design*. Beyond Architecture: New Intersections & Connections Methods: Agents of Change in Changing Paradigms. Scientific, Technological, Strategic, Intuitive, and Pragmatic, 229–237. Retrieved from https://www.arcc-repository.org/repository/ARCC2014_UHawaii%20Conference%20Proceedings.pdf.

Bolin, R. (1982). *Long-term family recovery from disaster*. Boulder: Institute of Behavioral Science, University of Colorado.

Bolin, R., & Stanford, L. (1991). Shelter, housing, and recovery: A comparison of US disasters. *Disasters, 5*(1), 24–34.

Cuny, F. C. (1983). *Disasters and development*. New York, NY and Oxford, UK: Oxford University Press.

Harrell, E. B., & Michael, J. Z. (2000). Including informal organizations in disaster planning: Development of a range-of-type measure. *Tulane Studies in Social Welfare, 21*(22), 61–83.

Kaiser Family Foundation. (2007). *Giving voice to the people of New Orleans: The Kaiser Post Katrina baseline survey (No. 7631)*. Retrieved June 21, 2010 from http://www.kff.org/kaiserpolls/upload/7631.pdf.

Karenga, D. (1992). Putting decision-making back into the hands of the people. *Appropriate Technology, 19*(3), 5–7.

Kincheloe, J. L., & McLaren, P. (2000). Rethinking critical theory and qualitative research. In N. K. Denzin & Y. S. Lincoln (Eds.), *Handbook of qualitative research* (2nd ed., pp. 279–313). Thousand Oaks, CA: Sage.

Lang, J. (1994). *Urban design: The American experience*. New York, NY: Van Nostrand Reinhold Company Inc.

Miculax, R. M., & Schramm, D. (1998). Earthquake housing reconstruction and rural development: Joyabaj, Quiché, Guatemala. In M. B. Anderson & P. J. Woodrow (Eds.), *Rising from the ashes: Development strategies in times of disaster* (pp. 225–240). Boulder, CO and San Francisco, CA: Westview Press.

Morrow, B. H., & Peacock, W. G. (1997). Disasters and social change: Hurricane Andrew and the reshaping of Miami. In W. G. Peacock, B. H. Morrow, & H. Gladwin (Eds.), *Hurricane Andrew: Ethnicity, gender, and the sociology of disasters* (pp. 226–242). New York, NY: Routledge.

Ozerdem, A. (2003). Disaster as manifestation of unresolved development challenges: The Marmara earthquake in Turkey. In M. Pelling (Ed.), *Natural disasters and development in a globalizing world* (pp. 199–213). London, UK: Routledge.

Peck, M. S. (1998). *The different drum: Community making and peace*. New York, NY: Simon & Schuster.

Rawal, V., & Desai, R. (1994). *Repairs and retrofitting: A manual for people of Marathwada*. Ahmedabad, India: Ahmedabad Study Action Group (ASAG).

Smith, G., & Wenger, D. (2006). Sustainable disaster recovery: Operationalizing an existing framework. In H. Rodriguez, E. Quarantelli, & R. Dynes (Eds.), *Handbook of disaster research* (pp. 234–257). New York: Springer.

Williams, H. (1977). On the formation of travel demand models and economic evaluation measures of user benefits. *Environment Planning, 9*, 285–344.

Wisner, B., Gaillard, J. C., & Kelman, I. (Eds.). (2012). *Handbook of hazards and disaster risk reduction*. Abingdon: Routledge.

Appendix

Port Bolivar Settlement Layout Along Broadway Avenue—Permeability Rubric Study

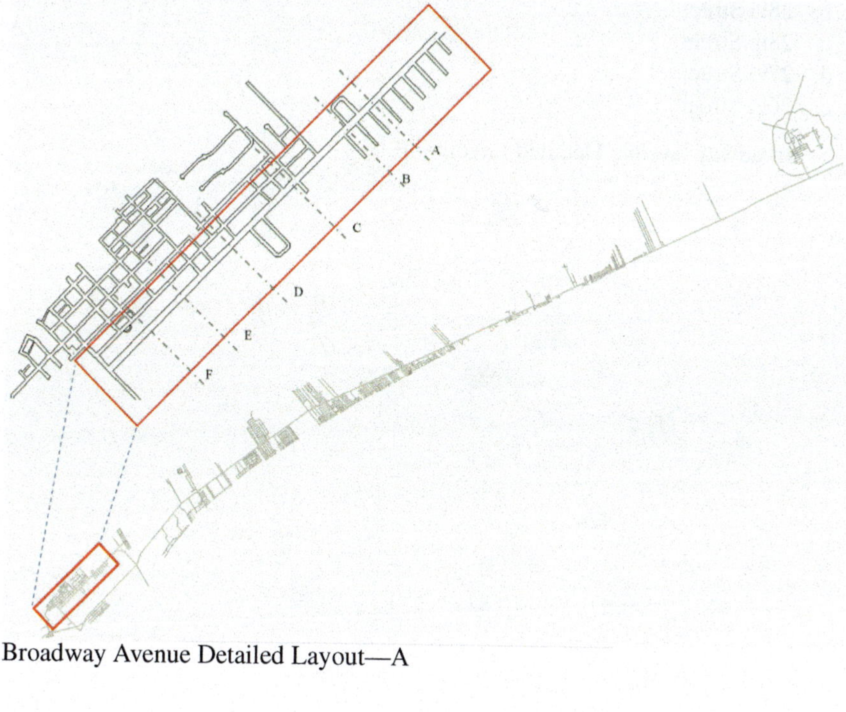

Broadway Avenue Detailed Layout—A

C. Subasinghe, *Rebuilding for Resilience: A Barrier Island Case*,
https://doi.org/10.1007/978-3-030-65532-7

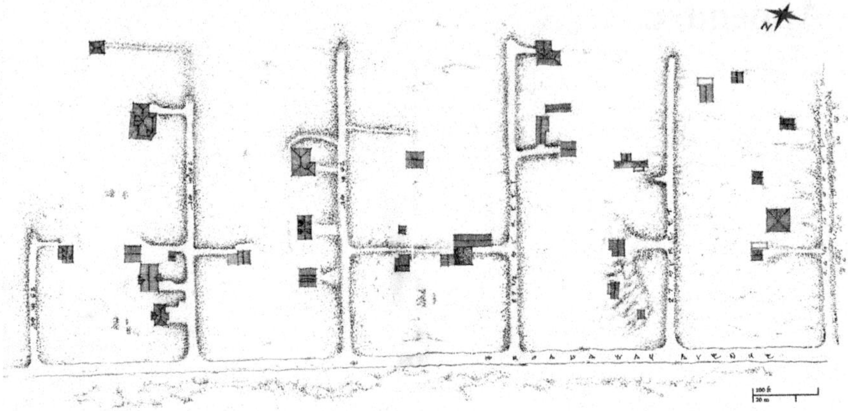

Layout-study showing properties at (from left)

a. 29th Street
b. 28½ Street
c. 28th Street
d. 27½ Street
e. 26½ Street

Broadway Avenue Detailed Layout—B

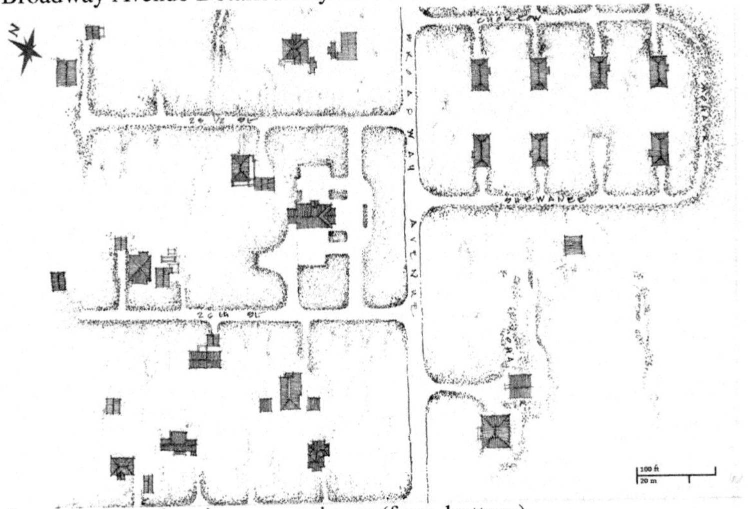

Layout-study showing properties at (from bottom)

a. 26th Street
b. Shewanee
c. 25½ Street
d. Choktaw

Broadway Avenue Detailed Layout—C

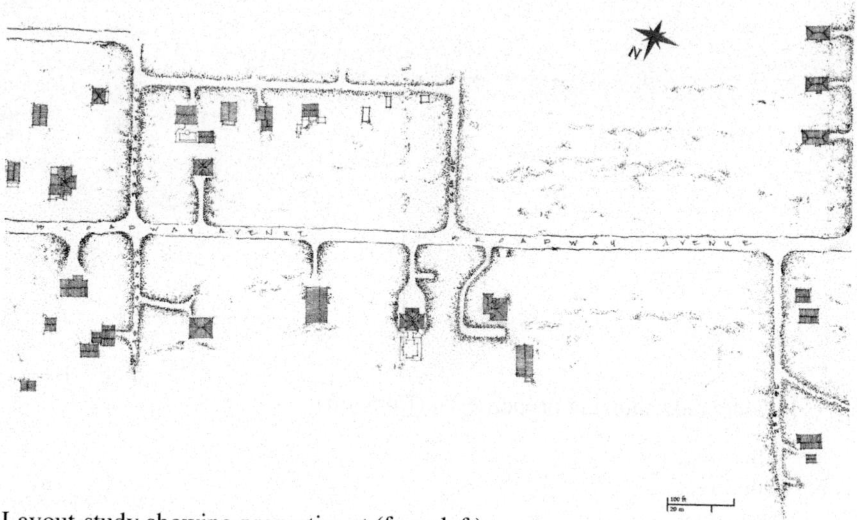

Layout-study showing properties at (from left)

a. 25th Street
b. 24th street
c. 23rd Street (North side of Broadway Avenue)/Ernest (South side of Broadway Avenue)

Broadway Avenue Detailed Layout—D

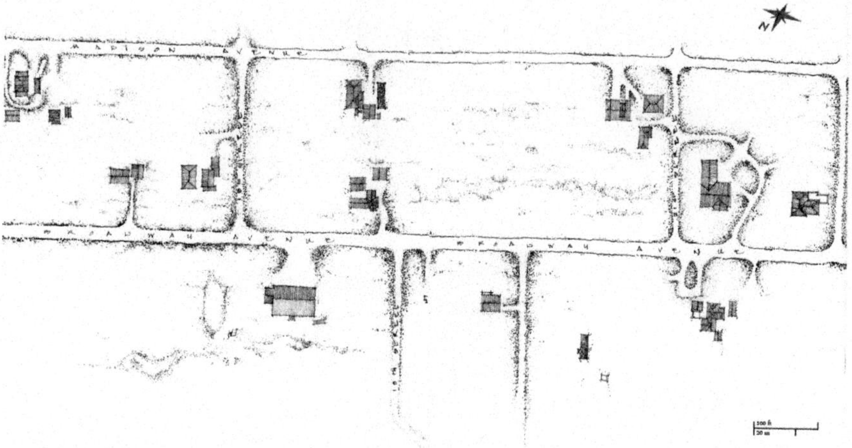

Layout-study showing properties at (from left)

a. 22nd Street
b. 21st Street
c. 20th Street
d. 19th Street
e. 18th Street

Broadway Avenue Detailed Layout—E

Layout-study showing properties at (from left)

a. 14th Street
b. 12th Street
c. 10th Street

Broadway Avenue Detailed Layout—F

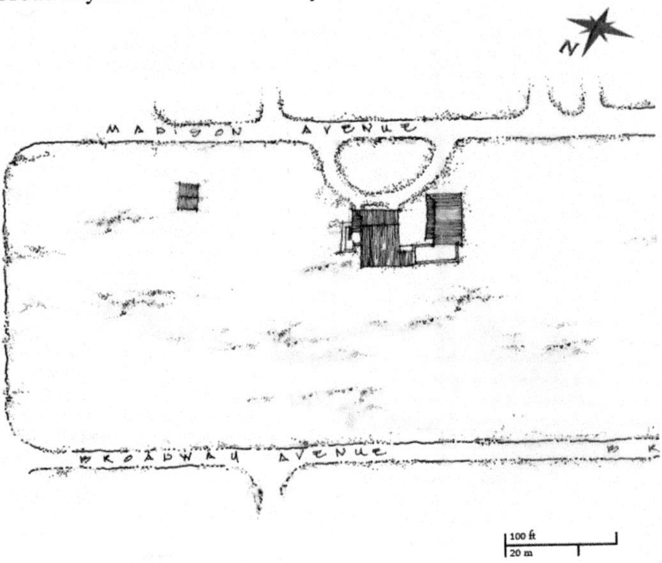

Layout-study showing properties at (from left)

a. 9th Street
b. 7th Street